Life on Other Worlds and How to Find It

Springer
London
Berlin
Heidelberg
New York
Barcelona
Hong Kong
Milan
Paris
Santa Clara
Singapore
Tokyo

Stuart Clark

Life on Other Worlds and How to Find It

Published in association with

Chichester, UK

Dr Stuart Clark
Director of Public Astronomy Education
University of Hertfordshire
UK

SPRINGER–PRAXIS BOOKS IN ASTRONOMY AND SPACE SCIENCES
SUBJECT *ADVISORY EDITOR*: John Mason B.Sc., Ph.D.

ISBN 1-85233-097-X Springer-Verlag Berlin Heidelberg New York

British Library Cataloguing-in-Publication Data
Clark, Stuart (Stuart G.)
    Life on other worlds and how to find it. – (Springer–Praxis
    books in astronomy and space sciences)
    1. Life on other planets   2. Exobiology   3. Space biology
    I. Title
    576.8′39

Library of Congress Cataloging-in-Publication Data
Clark, Stuart (Stuart G.)
    Life on other worlds and how to find it / Stuart Clark.
       p. cm. – (Springer–Praxis books in astronomy and space sciences)
    1. Life on other planets.   I. Title.   II. Series.
    QB54.C566 2000
    576.8′39–dc21   99-058201

Cover design: Jim Wilkie
Typesetting: BookEns Ltd, Royston, Herts., UK

Printed on acid-free paper supplied by Precision Publishing Papers Ltd, UK

For Libby – my very own alien

Table of contents

Foreword

Is there life beyond the Earth? Of all the questions that have been asked over the ages, this is probably the one to which we would most like the answer. If we are alone in the Universe, then we are indeed important. If not and there are countless civilisations, then we are truly insignificant.

In this book, Stuart Clark presents a masterly survey of the whole problem. First he provides a concise description of the make-up of the Universe, and he then discusses life in all its forms and asks some interesting questions.

What do we really mean by 'life' and can we place any credence in the aliens of *Star Trek*? Where can life evolve and how likely is it to do so? There is an overall survey of our neighbour worlds where there have been suggestions of possible life. We can certainly dismiss the brilliant canal-building Martians, but it is not so easy to dispose of the chances of life in the underground ocean of Jupiter's satellite Europa – always assuming that this ocean actually exists. Next we turn to planets of other stars, which must surely be present. It is significant that even as this book was going to press, British astronomers made the first visual sighting of an extrasolar planet orbiting a star more than 50 light years away from us.

Stuart Clark is an astronomer, but he also has a profound knowledge of biology and this is evident all the way through the text. He is not afraid to speculate, though he does so with commendable caution. The result is a book which is a mine of information but is also as entertaining as it is thought-provoking. It will appeal to the newcomer just as much as it will to the serious student.

When will we have a final answer? It is difficult to say. We may know more when we are able to handle rocks brought back from Mars, or when we establish whether or not the Europan ocean exists; or we may – just may – pick up a radio signal from a faraway civilisation. No doubt this book will need revising within a few years. Time will tell.

I read the book with intense enjoyment. In my view it is the best of its kind that I have seen, and it will have an honoured place on my shelf.

Patrick Moore
Selsey, 18 December 1999

Preface

'If you are interested in life in the Universe, this is the time to be alive.'
Lynn Harper, Astrobiology Scientist, NASA, September 1999

The above appeared in *New Scientist* on 18 September 1999. By coincidence this was the same week that I was due to hand over the finished manuscript of this book to the publishers. I took it as a good omen.

My involvement with searching for life on other worlds began one morning several years ago when I was in conversation with Professor James Hough. He mentioned that our long-time collaborator, Dr Jeremy Bailey, of the Anglo-Australian Observatory, had uncovered a possible link between the origin of life on Earth and the conditions we, as a group, had found inside certain star-forming regions.

As I and my colleagues began to read into the subject of the origin of life, I was thrust into a universe as alien to me as many find astronomy. Amino acids, nucleic acids, proteins – all exotic chemicals that I gradually came to appreciate as being the stuff of life on Earth. The more I became used to them and the way they interact, the more I saw parallels between biological systems and the physical systems of the Universe I was more accustomed to studying.

I developed a keen appreciation of life (and of the biologists who wrestle with its secrets) and understood for the first time that it is an integral part of the cosmos. At times, I even felt that I caught fleeting glimpses of the proverbial Big Picture. Yet, just as quickly as they appeared, they slipped away again.

In this book I have tried to convey my sense of the Universe and life's place within it. As you will soon come to realise, I believe that science is on the road to proving that life is as ubiquitous as stars in our Galaxy. The science of exobiology – as the study of life beyond the Earth is called – is now an established facet of modern science. No longer is the search for other life in the Universe a fringe discipline. It stands centre stage in the world's scientific research effort. Its time has truly come.

I have, therefore, used this book to present an overview of the science of exobiology for the general reader. No previous experience of life on other worlds is necessary – just a willingness to learn. I have even thrown in a few jokes along the way, to try to keep you amused ...

Stuart Clark
Hunsdon, 10 October 1999

Acknowledgements

This is the place to say 'thank-you' to all those people who have helped me either directly or indirectly while I have been writing this book. It is a constant source of amazement that so many seem to place so much faith in me. Thank you, one and all.

- Clive Horwood, Chairman of Praxis, who believed in this project from the beginning.
- John Mason, for his insight and encouragement.
- John Watson at Springer-Verlag, for his support, a great lunch and his musical stamina.
- David Anderson at Springer-Verlag for providing me with the opportunity to 'strut my stuff' in January 2000.
- Bob Marriott, for enduring the stress of having to edit this book in a style far more informal than that to which he is accustomed.
- Caroline Davidson and Alice Hunt for their undivided support and encouragement.
- My readers: John 'This is seriously *deep*' Atkinson, Julian Hiscox, and especially Jim Collett.
- David Axon, whose indomitable enthusiasm for astronomy led directly to my being appointed Director of Public Astronomy Education at the University of Hertfordshire.
- James Hough, for his crucial support in the above appointment.
- Chris Courtiour and Steven Young at *Astronomy Now*.
- My friends in the various amateur astronomical societies around the country.
- Special thanks to Don Tinkler (my biggest fan), Nik Syzmanek, Nick Hewitt, Mick Hurrell, Christine Dean, Sue McLintic, Christian Kay, Mat Irvine,

Steve Arnold, Frank Cliff, Jerry Workman and the attendees of my short courses for their loyal support.

- All those fine rock musicians who populate my CD collection and keep me company when I am writing. A special mention for Dream Theater – victim of my English spelling checker in my previous book, (an error which was not noticed by Bob Marriott, who, being a classical music fanatic, had never heard of them).
- For those moments when writer's block sets in: Phil Burrell – my guitar-yoda – who guides me on the path to harmonic enlightenment ...
- Finally, the amazing Nikki, who tore into this project with enthusiasm – and a large stock of red pens.

Chapter One

Test for echo

A s one of my favourite musicians once wrote: 'Test for echo – is there anybody out there?' Neil Peart, drummer and lyricist with Canadian rock group Rush, was referring to the creative path that his band was treading after a lengthy hiatus and he was wondering whether their audience had bothered to wait around. Being an intelligent and deeply thinking man, the application of the phrase to the search for extraterrestrial intelligence was not lost on him either.

I was instantly taken with the phrase. There is something primal about it and the uncertain feelings that lie behind its utterance. None of us – with the exception of a few hermits – wants to feel alone. For many, this feeling extends beyond family and friends and out into the cosmos. Astronomy leads us to believe that the Universe is so vast that we, on planet Earth, are nothing more than an insignificant mote. To me, this makes our existence sound pretty lonely. It is no wonder then, that many people contemplate whether other alien civilisations exist. Their hope is that, with the onward march of technology, one day – maybe long after their own lifetimes but one day nonetheless – humankind will be able to contact those other beings and the Universe will not seem such a lonely place after all. It is a comforting thought – but one that, for many, is based more on faith than on hard evidence.

Indeed, the belief in the existence of extraterrestrials has often seemed to me to be more like a religious conviction than one built on strong scientific principles. The best statement of this sentiment occurs in the film of Carl Sagan's book, *Contact*. Two of the characters say that if humankind is the only life in the Universe, then it seems like an awful waste of space. As much as the romantic in me shares this sentiment, it is not a basis for believing in ET's existence. In this book, I am going to try to be as scientific as possible. I will try to weigh the scientific evidence both for and against the existence of other life-forms in the Universe. There will be a lot of speculation; it is almost unavoidable, because there are still some rather large questions which remain to be answered about our basic knowledge of life, its origin, what falls within the definition of life, the number of planets in the Galaxy which could support life – indeed, whether life even needs a planet to live on. You get the picture. I will try, wherever possible, to make educated guesses and, where I cannot, I will clearly state that I am taking wild shots in the dark.

Let us begin this book with a brief discussion of the major assumptions that underpin the search for extraterrestrial intelligence (SETI). I am going to look at the assumptions very hard and be very critical. However, by the end of this book (as you will see) the act of being so critical will have led me to a remarkable realisation – that the search for extraterrestrial life and intelligence is one of the most important scientific tasks that can be undertaken. To find out why, keep turning the pages.

The research I have conducted to enable me to write this book has convinced me that extraterrestrial life is out there and that some of it is intelligent. All humankind has to do is look hard enough to find it. Perhaps I now understand why the phrase 'Test for echo' resonates so deeply within me.

PRINCIPLES AND ASSUMPTIONS

According to internationally renowned astrophysicist and science writer Paul Davies, there are three principles which must be invoked before it can be believed that other life forms will be found in the Universe:

> The Principle of Uniformity of Nature
> The Copernican Principle
> The Principle of Plenitude

Each one of these principles is actually implicitly assumed by most astronomers and cosmologists because, without them, the meaningful study of the Universe would be just about impossible.

Let us run through them one by one. The Principle of Uniformity of Nature basically states that the laws of nature are the same throughout the Universe as they are here on Earth. I think this one is on pretty safe ground. Astronomers and cosmologists, looking billions of light years into space, see events taking place that can be understood by applying the laws of physics. No matter where we look in space, we see galaxies and stars. If the laws of nature were different throughout space, we would expect to see phenomena that are completely inexplicable. Although I do not wish to sound complacent and imply that we understand everything in precise detail, I think we have certainly explained enough of the broad strokes of the Universe to be confident that the laws of nature are the same everywhere.

The Copernican Principle is tied up with the implications of the Polish astronomer Nicolaus Copernicus' book *De Revolutionibus Orbium Coelestium*, published in 1543. It was Copernicus who proposed that the motion of planets in Earth's night sky could be explained by placing the Earth and the other five known planets in orbit around the Sun. Copernicus thereby removed us from our assumed position at the centre of the Universe. This was a severe blow to the teachings of the Church and began the trend of divesting humankind of any position of privilege. American astronomer Harlow Shapley continued the 'Copernican revolution' by showing that our Solar System is not even located in the centre of the Galaxy.

The Copernican Principle is central to modern cosmology because, by stating that the Earth does not exist in a privileged position, we can be certain of studying a representative part of the Universe. If we were in a special location, we could never really know what a typical place in the Universe is like and so our cosmological theories would be hopelessly biased.

My feeling is that the ultimate vindication of the Copernican Principle would be if Earth, as a planet, were not even privileged by the emergence of life. If Copernican ideas can be extended to biology, then there is no reason to assume that life on Earth is a cosmic one-off. We will return to this point in a little while.

The final principle is that of Plenitude. The best way to think of this is that anything that *can* happen *will* happen. It is the Principle of Plenitude that lies at the heart of science's ability to make predictions. Picture the typical stereotype of a scientist: a middle-aged man with unruly white hair who slaves over a blackboard covered in abstruse mathematics. His writing becomes more and more hurried. Sweat beads his brow. With a flurry, he underlines his answer, shouts 'eureka!' and runs out of his office down to his laboratory (which is probably in his basement because he is an unacknowledged genius). Crashing through the door, he dons the obligatory white coat and begins to frantically assemble machinery from parts that have been lying around for years because he knew one day they would come in handy. Going without sleep for days and living on a diet of caffeine, he finally finishes his masterwork. By creating and focusing the correct physical forces he has realised that he can open a time portal from the present to a week last Tuesday. Throwing the switch, the temporal gateway opens. He boldly steps through and, in the nick of time, manages to rescue his pet tortoise, Horace, from that freak one-in-a-million accident involving a runaway steam roller ...

The point I am seeking to make (apart from the fact I will never make it in Hollywood as a sci-fi writer) is that, through mathematical descriptions of the Universe, science can show what is possible. By reproducing the necessary conditions, the possible becomes the actual. When the predictions are not realised in nature it is interpreted as meaning that a piece of our theory is wrong or is missing. Our understanding is not as complete as we thought and so back to the drawing board we go.

If I combine the Copernican Principle with that of Plenitude, then I am left with a startling prediction. According to the Copernican Principle there is nothing special about Earth. In other words, we should be able to find plenty of other Earth-like planets throughout the Galaxy. According to the Principle of Plenitude, given the right physical conditions, the possible is realised. Since Earth produced the correct physical conditions for life to appear, these two principles predict that we should expect life to be widespread throughout the Galaxy.

So, right from the outset, two of the deepest principles that astronomers believe in lead us to expect the existence of extraterrestrials.

Having accepted these three principles, we can proceed with the task of making assumptions. The Australian physicist David Blair has proposed five:

> The inevitability of life and technology
> Temporal mediocrity
> No superscience or hyperdrives
> Exploration by data exchange
> The 'Galactic Club'

I will briefly overview each one of these in turn because the evidence, both for and against these assumptions, will form the central theme of this book.

THE INEVITABILITY OF LIFE AND TECHNOLOGY

Life and technology – therefore, intelligent life and not just pondweed – are inevitable consequences of universal processes. Wow, what a statement! As we saw in the previous section, this assumption rests on the combination of the Copernican and the Plenitude Principles.

According to the Nobel prize-winning chemist Christian de Duve, life might indeed be a chemical inevitability. There are a number of notable scientists, mostly from the chemical and physical sciences, who are also swinging this way in their thinking. For what it is worth, I also think that life is basically nothing more than chemistry (as I will discuss in Chapters 4 and 5), so I think that given the correct conditions, life is inevitable. Biologists, in general, seem less optimistic. Perhaps their tempered reasoning can be understood by thinking of the tenets that, through their eyes, govern living things: the theories of evolution, all of which have sprung out of the work of Charles Darwin.

In 1859, Darwin published *On the Origin of Species by Means of Natural Selection or Preservation of Favoured Races in the Struggle for Life*. This seminal work was the product of years of work by Darwin, based on his expedition to the islands and coastal areas of South America. Most famously, the Galapagos Islands are inextricably linked to the nineteenth century scientist. His work on the distribution of living species across this area led him to believe that life could adapt to its surroundings. This flew in the face of contemporary thinking in which natural philosophers believed that the species inhabiting Earth were unchangeable and had all enjoyed a separate creation. Most people also believed that these creation events were the work of God.

Darwin's suggestion – that life-forms could be interrelated and could change with time – sent shock waves through conventional biology and through the Church, in much the same way as Copernicus had rocked fundamental scientific and theological belief with his work of 1543.

At the heart of life's ability to adapt was Darwin's theory of natural selection. It stated that it was possible for organisms to change slightly and randomly from generation to generation. Most of these changes were either of no importance or, even worse, detrimental to the organism. Every now and again, however, a chance mutation made the organism better suited to its environment and made it more likely to survive to maturity, reproduce and pass on these traits to future generations. In this way, the organism could evolve. Each step in the evolution was, however, the product of random chance.

Any random change to a complex 'machine', such as a living organism, is more likely to spoil things rather than make it work better. Imagine randomly cutting and then reconnecting a few wires in your video recorder whilst blindfolded. Would you expect it to work better by producing a sharper picture or by rewinding the tape faster? You would succeed only in invalidating the guarantee.

Life comes with no such guarantee. When mutations occur, the majority confer handicaps on the individuals which reduce their ability to live effectively in their environments, so they are more likely to fall prey to other animals before being able to reproduce. Imagine being born a mutant red frog in a jungle dominated by blue flowering plants. That is very bad news if your parents were hoping for grandchildren.

The random nature of mutations mean that many more will be detrimental rather than advantageous. This means that the probability of suffering an advantageous mutation must be very low. So, the adaptability of modern species to their environments is a product of evolutionary steps which have incredibly small probabilities. This is known to biologists as 'contingency', because each evolutionary step is thought to be contingent on the previous ones. It is the concept of contingency that lies at the heart of science fiction's most outlandish aliens. For example, contingency states that the chain of events which produced humans has such a low probability that if we were to rewind time and begin life on Earth again we would not expect it to follow the same random chain of events. In fact, in all reasonable probability, it would not follow anything remotely like the same chain of events and so the life forms which developed would be completely different from us. If the origin of life itself were the product of low-probability interactions between chemicals, then, the biologists argue, there is actually no basis for believing that life would begin again even if the conditions on the primitive Earth were precisely recreated.

Even if life did begin, there is no guarantee that it would become intelligent. After all, the mutations that led to intelligence must have been of low probability, so why should we expect it to happen again? For evidence of this, consider the dinosaurs. In terms of longevity, they are easily the most successful species so far ever to have lived. Their reign spanned 150 million years and they died out about 65 million years ago, to be followed by the rise of mammals. Compare that with the approximately 4 million years that the hominids (humans, our cave men and man-ape ancestors) have been around. Indeed humans themselves – *homo sapiens* – have been on Earth for only 300,000 years. Clearly, we are merely beginners at this 'world domination/top of the food chain' game. However, in all the time that dinosaurs ruled the Earth, they did not develop intelligence. Mammals, on the other hand, achieved it within 65 million years. I will explore this fascinating question in Chapter 9, when I consider the rise of intelligence on the Earth and whether it could happen on other planets.

TEMPORAL MEDIOCRITY

According to this assumption, human beings on the Earth exist at no special time. It is rather similar to the Copernican Principle, except that it seeks to divest humanity of any special place in time rather than in space. In more mathematical terms, temporal mediocrity can be expressed as meaning that the average time it takes an intelligent species to develop on a suitable planet is much less than the age of the Galaxy. If this is true, one can expect the Galaxy to be teaming with intelligent life.

The caveat in the definition is the phrase 'suitable planet'. Just what constitutes a suitable planet is an open question that we will tackle in

Chapter 6. I have already explained that if Darwinism is applicable to the origin of life, then an exact carbon-copy of Earth may not be suitable because of the contingency concept of low probabilities. On the other hand, how do we know that something quite different from Earth is unsuitable? Maybe it is only unsuitable for life as we know it (ouch! – we are only in Chapter 1 and the first *Star Trek* cliché has slipped in – and unfortunately, it will not be the last!)

Now we have really opened up Pandora's Box. In order to think about life in its most general form, we first have to define it – and this is where the problems begin. Life would appear to be a pretty recognisable quality but trying to pin down exactly what it is proves very difficult.

Life has a few fundamental traits that apply across the board. In all its forms it takes in food, converts this to energy and produces a waste product. This is not bad to start with. But think of a flame, which takes in 'food', converts it to energy and gives off a waste product. Clearly, however, a flame is not alive, so there must be something else. What about the ability to reproduce? This is a fundamental trait of life and is perhaps the strongest, most deeply ingrained one. Unfortunately, flames can also reproduce, otherwise starting a log fire with a match would be impossible.

We are forced to look deeper still. What about the fact that life is capable of adapting to its surroundings – the process of evolution through mutation? This ability is wrapped up in the molecule DNA – deoxyribonucleic acid. It is the molecule that exists in just about every cell in our body and carries the information to make our bodies. In fact, DNA is found in every living creature on Earth. The individual units of information carried by the DNA are known as genes and it is these that suffer the mutations and give us the ability to evolve. That sounds pretty fundamental to life.

Yet even mutability as a definition has serious drawbacks. A simple crystal can develop faults in its structure that are repeated as new crystal layers grow on top of it. In a sense these are mutations just as certainly as a changed gene. As I will discuss in Chapter 4, life cannot be pinned down to a simple snappy definition. Instead it is a collection of properties without any single defining trait. For now, however, let us content ourselves that DNA is important to life because of the information it carries.

I have been purposefully vague in the previous sentence because there is a point that I wish to make very forcefully. It is a common misconception that DNA is the key to life. *I* think it springs from the fact that here, on Earth, all life is based on DNA.

If we are saying that life is somehow wrapped up in the information carried by DNA, then we are not actually saying anything about DNA. The carrier of the information is relatively unimportant; it could be DNA, some other molecule or a paper bag for all we care. The important item is the information.

After all, a great book is not judged by the paper it is printed on but by the information contained within its pages.

We have become so blinkered in our organic chemistry, simply because all life is based on the same basic molecular design, that I believe it has closed our minds to considering other, alien alternatives. In fact, if contact is made with extraterrestrials then I believe that our first enquiry should be 'Please explain your biochemistry.' (To be honest, it was a toss-up between the biochemistry question and 'Can you recommend a good translator for my books?' I am certain that those of an overtly religious disposition would have a different first question.)

Romantic notions that aliens will be organically similar to us and so advanced that they will arrive with all sorts of medical knowledge (such as an umbrella cure for cancer) are pure fantasy. To me they echo the religious desire for salvation from a higher power, rather than the dispassionate, scientific consideration of extraterrestrials.

But what if the romantics are correct and the first aliens who land on Earth are based on DNA? The ideas of Darwinian contingency will be blown completely out of the water. Evolution in two completely separate places would have produced the same life-giving molecule. It will almost certainly be seen as proof of the existence of a godhead by some, because it smacks of design and purpose. Others will claim that life must have started on one planet and then somehow have been transplanted to another. This latter idea is known as the 'panspermia hypothesis', which I will discuss further in Chapter 5.

NO SUPERSCIENCE OR HYPERDRIVES

'Scotty, I need warp speed in three minutes or we're all dead ...' Well, according to the assumption of no superscience or hyperdrives, everyone's favourite starship Captain, James T. Kirk, is going to have to wait an awful lot longer than three minutes for Scotty to perform his engineering magic.

It is very easy to think that technological advancement is inexorably progressing, as this is the way it seems from our perspective in history. In fact, when one thinks of the advances that have had profound effects on our civilisation, it seems that the pace of technological advancement is exponential in nature – beginning slowly before gaining momentum and accelerating quickly (see the illustration on p. 9).

But can this kind of growth be sustained? There are those who believe that physics can now explain the Universe so well that there are no fundamental concepts that remain undiscovered. If this view is correct, then we confront the concept of a scientific glass ceiling. Despite the fact that our imaginations can

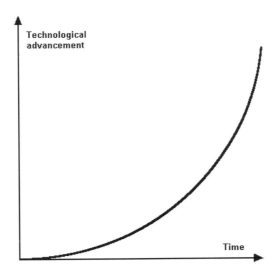

The exponential curve of technological advancement. This would be the pattern of advancement if physics is a never ending treasure chest of new ideas and possibilities.

picture starships forging through the void, the laws of physics forbid it. If it is true that such a ceiling exists and that we are close to it, then we would expect the rate of technological advancement to tail off as we approach closer and closer to exploring all the possibilities our Universe has to offer. In this case, the graph of technical achievement with time would look sigmoidal in shape (see the illustration on p. 10).

SETI assumes that aliens will use radio waves to transmit messages into space. Imagine our chagrin if, after years of effort listening to radio frequencies for alien messages, a spacecraft lands on the lawn at the observatory at Jodrell Bank. Then, when the director of the observatory shows the aliens the sophisticated receiving equipment that has been employed by the SETI searches, they remarked 'How quaint; we gave those up centuries ago.'

After all, we would not tie a piece of string to a tin can and launch one end into space if we wanted to communicate with astronauts on the Moon. If superscience methods of communication are within the grasp of a civilisation within a reasonable time of its discovery of radio waves, then SETI at radio frequencies is doomed to failure. Hence, a SETI scientist must believe in the no-superscience assumption, or believe that aliens will transmit purposely on radio so that they can talk to technologically backward planets.

Personally, I believe that there is no ceiling, or that if there is, it is so far away that it is not yet in sight. My view is encouraged by the historical progression of gravitational theory. When Isaac Newton published his universal

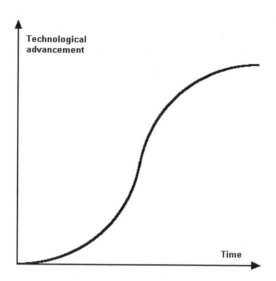

The sigmoidal curve of technological advancement. This is the shape of technological advancement if a ceiling to our physical knowledge exists. Even if this is the correct shape of the curve, we do not know whether we are near the top of the sharp rise – indicating that the speed of advancement will soon begin to slow down – or near the start of the sharp rise – indicating that we 'ain't seen nothing yet'!

law of gravitation in 1687, it seemed that one of the most fundamental truths of the Universe had been discovered. It resulted in the scientific revolution that led to the view of the 'clockwork Universe' and dispensed with God's day-to-day help in Earthly transactions. Arrows moved only according to Newton's laws, not as a result of the kindly intervention of God's hand. So it appeared that no amount of praying would make your arrow fly any straighter than your own skill dictated.

A tiny niggle in Newton's triumph was that the innermost planet of the Solar System did not move according to the way that Newton's law prescribed. Was it perhaps God's hand after all, perturbing the planet to keep it habitable for Mercurians?

Remember what was said in the discussion of the Principle of Plenitude – that when science cannot predict something it means that a deeper theory is required. For the explanation of gravity, the deeper theory was provided by Albert Einstein's theory of general relativity in 1915. This correctly explained the motion of an object that found itself deep in a relatively strong gravitational field, such as tiny Mercury orbiting so close to the gigantic Sun. So, out of this small inconsistency in Newtonian theory has sprung the modern understanding of black holes, active galaxies and the origin of the Universe, to name but three.

I do not think, therefore, that any of the tiny inconsistencies in our current view of the Universe can be dismissed as trivial. Their solutions may not simply be the icing on the cake of modern theories but, rather, doorways to wholly different and deeper theories of the Universe.

I fully acknowledge that the field of physics, as it stands at the moment, does not allow the passage of material objects through the spacetime continuum at velocities faster than the speed of light. Yet those same laws allow the spacetime continuum itself to move faster than light. In fact, the resolution of two major cosmological problems – the horizon problem and the flatness problem – demands it.

I think the biggest 'niggles' in modern physics lie in the quantum theory. According to this theory it should be expected that gravity itself is carried on tiny particles called gravitons. As yet, no one has developed a convincing theory of quantum gravitation. Who knows what possibilities it would unlock? General relativity itself does not preclude conditions that would make short cuts through time and space possible. These are concepts that have found celebrity in science fiction under the name of 'worm holes'.

Of course, it may be impossible to formulate a quantum theory of gravity because there is no such thing. Hence a convincing mathematical description of a quantum particle to carry the force cannot be formulated, because gravitons simply do not exist. Humankind would then have hit one of those ceilings to physical knowledge and Newton and Einstein would have nailed down the theories pretty much completely.

The other possibility is that a quantum theory of gravity requires such a tremendous intellect to grasp, that we are very far away from developing the conceptual and mathematical skills to reduce it to our level of understanding.

It is always important to remember that, just because it is possible to ask a question, there is no guarantee that it can be answered. Indeed, the quantum theory of physics tells us that at the level of subatomic particles there are some questions that cannot be answered. There are certain conditions in the Universe which make it appear as if particles can share information faster than the speed of light. What is the tie that binds these particles together? Perhaps when physicists understand that, they will be capable of finding ways around the cosmic light-speed limit. In Chapter 11 I will present a highly speculative idea about instantaneous communication based on the poorly understood quantum phenomenon of entanglement. Although this particular idea itself may be totally unworkable, I maintain that science continues to open new possibilities all around us.

EXPLORATION BY DATA EXCHANGE AND THE 'GALACTIC CLUB'

I shall deal with the final two assumptions together. Really, they follow directly on from the no-superscience assumption. If there is no way for intelligent beings to physically explore the Universe, then it is assumed that they will try to make contact with other beings on other planets and simply ask each other questions.

Remember I said that the most important question would be about biochemistry? Well, that is my thirst for knowledge talking. The assumption of exploration by data exchange is built around that thirst. Closely linked to that is this 'Galactic Club' assumption which postulates the existence of a network of planetary civilisations that exchange data. Because new members bring new data and knowledge, they are obviously welcome. In fact, the 'club' may even actively seek out new members by beaming messages towards candidate planets.

This is all starting to get a little too far out for me. The reasoning behind this Galactic Club implicitly assumes that aliens are as curious about the Universe as we are. Although to humans it sounds like a good idea to explore by data exchange, if we can do it no other way, an alien, with different physiology, biochemistry, background and so on, would almost certainly think very differently. So the real flaw in this argument seems to me to be the implicit assumption that aliens will think in ways that are similar to our own.

There is a particular branch of behavioural science on Earth that tries to understand the way in which animals think. In the case of a cephalopod – an octopus, for example – it is certainly different from the way a mammal thinks. Of course, most men in the world claim not to understand how women think and they are the same species! So how could we even begin to second-guess the way in which a hypothetical alien exercises its grey matter? – if its matter is even grey.

I therefore think that the final two assumptions are on rocky ground and the superscience assumption needs investigating a lot more carefully.

LIFE ON EARTH

Given all the things we do not yet know and have had to make assumptions about in the previous pages, it is at least comforting to know that we have a pretty good grasp of what life is all about here on Earth – right? Well, I'm not so sure about that now either. Within the last few years, biologists have discovered living organisms that exist within solid rock, many kilometres underneath the surface of the Earth. Add this to the lifeforms that have been found living in high-temperature water, in strong saline solutions, or even in environments dominated by sulphur and you have a seemingly unearthly zoo right here on our own planet. There is so much activity and interest in these life-forms at the moment that a new journal exclusively dedicated to extremophiles, (as these-life forms are called) has appeared on the shelves. It is called *Extremophiles: Life Under Extreme Conditions* and is published by Springer-Verlag. Although the title implies it, it does not include research on authors and publishers with book deadlines looming.

Interest in extremophiles has suddenly ballooned, because it now appears as if the origin of life itself may have taken place in extreme conditions. The discovery and study of these bizarre life-forms may ultimately reveal just how life processes began on our world somewhere between 3.5 and 4 billion years ago.

Interest in extraterrestrial life has also been boosted recently by further exploration of the Solar System. The Galileo spaceprobe has shown that there is almost certainly an ocean of liquid water underneath the icy crust of one of Jupiter's moons, Europa. Some scientists hope that microbial life may be found in the ocean. There are already designs for spaceprobes to go looking for life on Europa.

A few years ago the world was stunned by the announcement of fossil remains of martian bacteria in a meteorite found in Antarctica. The debate rages – and I really mean rages – over whether or not the findings are ancient martian bacteria. NASA is placing a huge emphasis on the exploration of Mars and it is specifically looking for the remains of life or for existing microbial life which may just be hanging on, by the skin of its proverbial teeth, under the hostile conditions that now prevail on the red planet.

With this wealth of new interest and information, it is time to re-examine all of the assumptions behind SETI, taking as full an account as possible of the explosion of new ideas that have invaded science over the past few years.

Chapter Two

The universal stage

For the purposes of this book, the entire Universe is the stage and so before we think about the possibilities of life developing on that stage, it is important to look at the scenery. With this in mind, the question we wish to examine first is: how soon could human life have developed in the Universe? This will (in some ways at least) test our assumption of temporal mediocrity.

The best estimates for the age of the Universe place it somewhere between 10 and 15 billion years old. The age of the Solar System is 4.6 billion years. Fossil evidence suggests that primitive microbial life was present on Earth 3.5 billion years ago and probably earlier. Do these facts increase the belief that life on Earth has developed at no special time? Let me start right at the beginning of time and work forwards until we have the answer.

The very beginning of the Universe was a mysterious time – of which cosmologists endeavour to make sense. The hot Big Bang is the conventional theory for the origin of the Universe. It is usually accompanied by the add-on extra: inflationary theory. Like a software bundle that is given away with a new computer, inflation adds versatility to the Big Bang by helping to explain how and why celestial objects are as they are today. For example, it solves the horizon and flatness problems mentioned in Chapter 1.

The beauty of the Big Bang is that when the Universe was very young it was also very hot. This makes life a lot simpler for a physicist. Let me explain. Imagine a roulette wheel. The roulette ball is placed in the wheel and given a high initial energy by the croupier's flick of a wrist. This makes the ball roll around the top of the wheel. As it loses energy it settles towards the centre of the wheel, where it falls into one of a number of rotating slots. Not being a gambler, I have no idea how many slots there are but each slot represents a number. We can think of this number as being the final, low-energy state of the roulette ball. There are many low-energy states that the ball can land in, all of which are different and distinct. Even adjacent states are completely different from each other. To prove just how different, imagine placing your entire month's salary on lucky number seven and watching as the roulette ball falls into slot number eight. Although this is a number only one greater than yours, that is little compensation as you watch your money being dragged across the baize, away from you.

So even apparently close states are separate and distinct but the high-energy roulette ball spinning around the top of the wheel is a very different story. At this point in a roulette ball's history, it is indistinguishable from any other spin of the wheel. In this initial state, all roulette balls behave in the same way, which is what draws people back to the table – the unbiased chance of winning. Only when the roulette balls lose energy do their properties – and by that I mean the number they represent – become different.

In the high-temperature, high-energy Universe of long ago, all constituents of matter and energy behaved in a very similar way, rather like high-energy roulette balls – rushing around, crashing into each other and behaving in a remarkably similar fashion to the idealised way in which physicists imagine a perfect gas behaves. They envisage microscopic billiard ball-like particles bouncing off each other, transferring energy to one another and keeping everything as homogenised and equal as possible. In fact, the whole Universe was a virtually uniform mixture of similar particles and energy. It was a nice, orderly place – even simple.

As the energy of space dropped, however, things changed. One of the jobs of a cosmologist is to follow the way in which the Universe's constituents lost that energy and became separate, distinct entities that could then go on to interact in a number of interesting and ever more complicated ways.

There was so much energy in the early Universe that matter was an ephemeral thing. It fluctuated back and forth between becoming pure energy and then spontaneously forming back into matter. The Universe's capacity for this back-and-forth creation and destruction of matter is embodied in one of the most famous equations in physics: Einstein's $E = mc^2$. E is the amount of energy needed to create a particle of matter with mass, m. Sometimes E is known as the mass energy. The other term in the equation, c, is the speed of light (300 million metres per second) and is needed to relate a mass in kilograms into an energy, measured in units called joules.

This implies that any particle can spontaneously form from a tiny volume of space which packs in more energy than the particle's mass energy. In reality there is a slight complication. In order to allow matter to turn back into energy, a particle of antimatter must also be created. A particle of antimatter is almost identical to its matter particle counterpart in that it possesses the same mass. Its biggest difference is that its electrical charge is opposite in sign. There are some other differences too but we shall leave the particle physicists to worry about those. The important point is that, if a matter particle collides with its antimatter counterpart, they will annihilate one another and convert all that pent-up mass energy to pure energy.

So, in order to keep the Universe's double entry bookkeeping system of matter and energy balanced, whenever a particle of matter is created, its antimatter counterpart also spontaneously comes into existence. Thus, the amount of energy required to spontaneously form a particle of matter is doubled.

MATTER IN THE UNIVERSE

If there really were this perfectly balanced creation and destruction of matter and antimatter, then a question springs immediately to mind. How is it that the Universe is made of matter? Surely it should have been annihilated by the antimatter that was created with it.

The Universe of particle physics is very complex. In fact, many of the reactions that take place are not governed by hard and fast rules – only statistical probabilities that such-and-such will happen. In the high-energy environment of the early Universe, not just the obvious particles – such as those that make up atoms – were capable of being created. In fact, many exotic particles, which are long gone in the Universe of today, were capable of being created. These could spontaneously decay into other particles. In the vast majority of cases these resulted in matter–antimatter pairs but about one in every billion reactions resulted in a particle of matter being created without its antimatter counterpart. This 'tiny' residue is what makes up every galaxy, every star, every planet, (every ET?) in the Universe, not to mention all of us.

As the energy of space fell, so particles of matter 'froze out'. This meant that there was no longer enough energy to make them spontaneously and so the total matter content of the Universe became fixed. The main particles of interest to us are those which form atoms: protons, neutrons and electrons. Protons and neutrons are what comprise the nucleus of an atom. The number of protons determines the chemical species of the atom. Each distinct chemical species in known as an element. Elements include hydrogen, oxygen, silicon, iron and uranium, to name only a few. The number of neutrons in an atomic nucleus determines which isotope it is. Some isotopes are radioactive.

Protons are positively charged, whilst neutrons are neutral. This means that the nucleus of an atom is always positively charged. The negatively charged electrons orbit the nucleus and, because there are the same number of electrons in an atom as protons, the atom itself is electrically neutral. Electrons can be lost from atoms turning them into positively charged ions.

If electrons are electrically opposite to protons, does this mean that electrons are the antimatter counterpart of protons? This is a common misconception and, in fact, physicists themselves courted the idea several decades ago. The answer is 'no', because an electron is a fundamentally different particle from a proton. It may have an opposite charge but it certainly does not have the same mass. It is much lighter.

The atom is a fundamental building block. After its first flush of youth (the first second or less) the Universe contained a sea of free-floating protons, neutrons and electrons. A single proton can, in principle, capture a single electron into an orbit about it. This forms the chemical element we call hydrogen. So a hydrogen nucleus is, in fact, just a single proton. The matter content of the very early Universe was, therefore, 50% hydrogen nuclei with almost as many neutrons thrown into the mix for good measure. Many of those neutrons rapidly decayed, each turning into a proton and an electron.

About one minute after the Big Bang, the Universe entered a period in which the energy of space allowed protons to join together with the remaining neutrons and other protons. After four minutes of this, the activity ceased, because the energy dropped still further, with the result that 75% of the protons remained untouched as hydrogen and the other 25% had combined with the neutrons to produce helium nuclei. These each contain two protons and two neutrons. Hydrogen and helium are the two simplest chemical elements.

The Big Bang seeded the Universe with matter made of these two simple elements in the overwhelming proportions of 75% hydrogen and 25% helium. All this amazing activity was made possible because energy was falling everywhere, allowing particles to behave in their own separate and distinct ways. So just how did the Universe lose energy?

THE FALLING ENERGY OF THE UNIVERSE

A roulette ball dissipates its energy mainly as friction with the roulette wheel but for the Universe the dissipation of energy is rather more difficult. Because the Universe is everything, it is a closed system. This is another idealised physical concept, which means that the system under study cannot communicate with the outside world. Indeed, in the case of the Universe there is no 'outside world' because the Universe *is* everything. It is the ultimate closed system.

'So how did it lose energy?!' I hear you cry. Well, in fact, it did not really lose energy at all. 'So how did its constituents assume lower energy states?' Here is the true cunning in the tale. The Universe has one overriding property: it is expanding. As it expands, the energy it possesses becomes spread over a larger and larger volume. Thus, although the total energy of the Universe remains the same, the energy density decreases.

Before proceeding any further, let us think about what energy actually is. In a standard physics textbook you will read that energy allows objects to perform work. Most of us will be familiar with the concept of energy in terms of children. Two-year-olds often have it in excessive quantities, whilst most teenagers seem to have a deficit before noon. It seems to me that there is also some bizarre (and as yet unacknowledged by the scientific community) law of physics that allows children to steal away your energy and use it, leaving you drained and craving sleep. But I digress. Thankfully, the laws of the physical Universe are not so unprincipled, as we shall see.

One of the most important forms of energy in the Universe is electromagnetic energy. This type of energy is carried around at the speed of light by electromagnetic radiation and, when it collides with a particle of matter, allows the particle to absorb energy and do work. That work may be to move or to interact with another particle, or simply to release the energy – which it has so recently absorbed – in a new direction. Light is a particularly familiar form of electromagnetic energy.

In modern physics there are two viewpoints – known as 'models' – to explain electromagnetic energy. Neither is fully capable of detailing everything about it and yet, between them, they fill in each other's gaps. So both viewpoints remain equally valid. It is up to the physicist to choose which model to use in the context that is being studied.

In terms of light propagation and the way in which rays of light interact with one another, it is best to think of electromagnetic rays as waves. In this particular context, light is a periodic variation in the electromagnetic field of space that looks like a mathematician's sine curve and carries energy (see the illustration on p. 19). The behaviour of waves perfectly explains phenomena such as diffraction and interference, which are observed to be characteristics of light rays. Being a wave, it will have the interrelated properties of wavelength and frequency. These quantities are detected by our eyes and, in our brains, are interpreted as colours. Red light is very roughly double the wavelength of violet light. All the other colours of the rainbow fall somewhere in between. The electromagnetic spectrum of light spans between 350 nanometres (nm) and 700 nm (a nanometre is one billionth of a metre). Electromagnetic rays above or below (or, as it is sometimes referred to, longward or shortward of) these boundaries are not light but are other types of electromagnetic

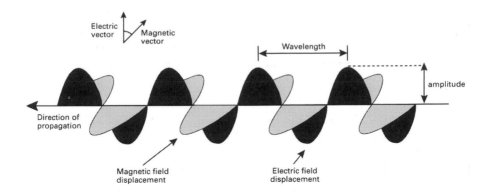

Electromagnetic radiation. A ray of electromagnetic radiation is composed of two sine curves. One is in the electric field of space and the other, at right angles, is in the magnetic field of space. Energy is transported in the direction that the wave travels.

radiation. For example, longward of the visible spectrum is the infrared, followed by the microwaves and then the radio waves. Shortward brings us first to the ultraviolet and then to the X-rays and gamma-rays.

Light interacting with matter is another story. The wave nature is then inadequate to explain what happens. When light travels through a cloud of gas, the way in which it interacts with matter is highlighted. Looking directly at a light source that gives off light because it is hot – a light bulb, for instance – allows a continuous spectrum to be seen with all the colours from red, orange, yellow, green, blue, indigo and violet being present. The best way to prove this is to pass the light through a prism, thus splitting it into its constituent wavelengths. Water droplets in the atmosphere do this to sunlight, causing a rainbow to appear. The compensation for a wet summer is the number of spectacular rainbows created!

If the same is done by instead looking at the light source through the cloud of gas, a pattern of dark lines will be superimposed on the continuous spectrum. This is known as an absorption spectrum because the gas cloud has somehow removed certain, very specific, wavelengths of the light from your line of sight. Looking at the same gas cloud, without the light source in the line of sight, shows a pattern of bright lines with no continuous spectrum. The bright lines correspond to the wavelengths of lights absorbed by the gas from the continuous spectrum. So this emission spectrum represents light that has been absorbed by the gas and then re-radiated.

Explaining this kind of behaviour – in which only certain wavelengths are absorbed and re-emitted by atoms of matter in the gas – is impossible when thinking of light as a wave motion. However, if light is thought of as a particle, it can be envisioned that the gas atoms 'cherry pick' from the light, absorbing only those wavelengths of light that they particularly like.

In the particle view of light, instead of energy being transported in a wave, each particle behaves like a tiny packet of energy. These particles are known as photons. Because a photon cannot have a wavelength but can carry an amount of energy, it is this which defines the colour seen when the photons strikes our eyes. Violet-light photons carry about twice as much energy as red-light photons. The energy of a photon decreases beyond the red, becoming smaller and smaller through the infrared, microwaves and radio waves. Conversely, the opposite happens through the ultraviolet, X-rays and gamma-rays, with an increase in the energy carried by the photons.

Having two complementary, yet completely irreconcilable, views on the nature of light is known as wave–particle duality (see the illustration below). In fact, physicists have become so accustomed to this concept of wave–particle duality that they are often blasé about mixing their metaphors. For example: double the energy of a photon and you halve its wavelength. It is verbal short-hand, or lazy use of language – depending on your point of view. But beware, it lurks everywhere!

Wave–particle duality. This difficult concept can be approximated by the classic 'two faces or a vase' diagram. It is usual to be able to see either two faces or a vase but it is impossible to focus on both at the same time. The same kind of thinking must be applied to electromagnetic radiation when thinking of it as either a wave or a particle.

To return to our gas cloud example. Something about the gas atoms must make them susceptible to absorbing only certain photons (that is, wavelengths of light). If you change the type of gas in the cloud, the wavelengths that are absorbed also change. So it is something about the structure of the atoms, making up the gas, that causes the absorption.

Remember that electrons orbit the centre of the atom. In the case of a hydrogen atom, one electron orbits the nucleus. It is possible for this electron to orbit in any one of a number of different orbits, known as energy levels. It can jump between energy levels (orbits) if it receives exactly the right amount of energy to do so. Just too much or just too little is no good; only the right amount will do. When a photon carrying just the right amount of energy collides with the electron, it absorbs that photon and jumps to the new energy level. Being a fickle thing, the electron will soon jump back to its original state and release the energy. This will again be carried on a photon but one that is released in any direction and almost certainly not in the direction that it was first travelling.

Since the energy carried by a photon dictates the wavelength of light it represents (I warned you about mixing wave–particle duality metaphors), it is the electronic structure of the individual gas atoms that gives rise to the pattern of lines in absorption or emission spectra.

In order to determine where the energy levels will be found, electrons must be treated not like particles but as – you've guessed it – waves. Instead of rushing around the nucleus like a minuscule planet on amphetamines, the electron must be thought of as a wave, joined in a ring, oscillating happily. Only those orbits in which a whole number of wavelengths can fit around the nucleus are allowed. So electrons experience wave–particle duality as well. No wonder they get on so well with photons.

The precise arrangement of energy levels around an atomic nucleus (and hence the pattern of spectral lines it produces) is determined by a number of different things, including the number of protons in the nucleus. Hence, by recognising the spectral line patterns, astronomers can determine the type of chemical elements present in celestial objects.

How does all this link in with the expanding Universe? Energy is carried by light and the other forms of electromagnetic radiation that stretch through space in waves. As space expands, so these waves become stretched, increasing their wavelengths. For visible light, this means that light is shifted towards the red end of the spectrum; hence astronomers term this stretching phenomenon 'redshift'.

Redshift can be most easily seen in the spectral lines of very distant objects. For example, a type of galaxy known as a quasar has a very bright centre which can be seen for an incredible distance across the Universe. Although astronomers can still recognise the spectral line patterns of familiar chemical elements, the patterns have been stretched and placed in unusual – but always longward from normal – positions in the spectra of these objects. This is exactly the kind of behaviour that would be expected if the light from these distant quasars was being redshifted. As a result, astronomers feel confident that space is expanding.

The leftover energy of the Big Bang is also carried on photons of electromagnetic radiation. In the initial, high-energy state of the Universe, these photons each carried enormous amounts of energy. This is what allowed them to spontaneously form matter–antimatter pairs. Because of wave–particle duality, however, these photons also behave like waves of energy spread across space. As the Universe expands, so these waves are redshifted and the photons associated with them drop in energy. Hence, the energy of the Universe appears to fall.

BEWARE THE SECOND LAW OF THERMODYNAMICS

It is time to introduce a concept that will return to haunt us again and again throughout this book: the second law of thermodynamics. This simple, unassuming name hides one of the most insidious and deadly laws of nature. It is this law that, in the end, will get us all.

Thermodynamics is the study of how objects transfer energy, allowing work to be done. It allows us to make a specific definition of heat. It is the amount of energy transferred between two bodies that happen to be at different temperatures. Heat is therefore a type of energy. Temperature is a property displayed by every object which is linked to the amount of energy it can potentially transfer to other objects. In any thermal reaction (one involving heat) energy is transferred from the object with the higher temperature to the object with the lower temperature. So in the early Universe, the frequent number of collisions between particles of matter and energy ensured that energy was shared around in a completely equitable fashion, so everything possessed the same temperature. This process is known as thermalisation and once achieved, the system is said to be in thermal equilibrium.

The expansion of the Universe and the leaching away of the photon energy this caused, kept the Universe from maintaining perfect thermal equilibrium.

The first law of thermodynamics can be thought of as stating that the total energy of a closed system remains the same at all times. Wrapped up in the first law is the concept that energy can be neither created nor destroyed. All that can happen to it is that one type of energy is turned into another.

'Ah-ha!', the inventive part of your brain says. 'All I need to do is to build a machine in which one type of energy is converted into another type and then convert it back again and I have a perpetual motion machine! Now, where's the telephone number of the patent office ...' Slow down! Many have already tried and all have failed. Why? Because any real system will lose energy, one way or another, to its surroundings and hence cannot be the physicists' ideal closed system. To keep the machine going, energy must be replaced to compensate for the energy which has been lost.

The idea of energy flow leads us to the second law of thermodynamics. This is a monster of a concept. In its simplest form (we will become more sophisticated as the book progresses) the best way to think of it is that heat cannot flow from a colder body into a hotter body. In other words, this is why your bath goes cold. The hot water cannot be supplied with heat from the cooler air of the bath room. We all intuitively understand this aspect of the second law from everyday experience but most of us simply do not know what it is called (apart from common sense). For example, how many of us say 'I am just waiting for the second law of thermodynamics to cool my cup of tea sufficiently so that a second application of the same law does not cause me to scald my mouth.'

It is not just bath-tubs and cups of tea, or even humans with warm bed-fellows on cold winter nights, that exploit the second law of thermodynamics. It controls the direction of energy flow right down to atomic scales too. So if we think of life as a chemical machine, it is obvious that the second law of thermodynamics is essential to all living things. We will return to this idea in Chapters 4 and 5.

Once objects reach thermal equilibrium – that is, they are at the same temperature – the transfer of heat stops. Again, this is a consequence of the thermodynamic second law. Heat can flow only from a hotter to a cooler object. If both objects are at the same temperature, no heat can be transferred. So, the second law of thermodynamics can be seen as driving the Universe inexorably into a position of thermal equilibrium.

Thankfully, it has a jolly long way to go. Even a casual glance at the night sky shows that the Universe is full of stars. These are in very obvious thermal disequilibrium with their surroundings, otherwise they would not be radiating energy into space. It is interesting and crucial to note here that a large fraction of life on Earth has developed to exploit the outpouring of energy from the Sun.

We hear a lot about humankind taking to starships and escaping the death of the Sun in a few billion years time. In fact, even if humankind can escape the eventual demise of the Sun by migrating to another solar system, long-term survival is not guaranteed nor, perhaps, even very likely. In the incredibly far-distant future, the constituents of the Universe will approach thermal equilibrium. This will rob the Universe of energy sources and, without those, life will not be able to carry on. So the second law of thermodynamics is humankind's ultimate enemy. This is why I called it 'insidious'.

WHAT WENT WRONG?

What seems very interesting to me is that when the Universe was born it was in an homogeneous state in which matter was spread evenly throughout space and thermal equilibrium was just about established. From such simple, even humble, beginnings springs the complex cosmos we have inherited today. But how? What went 'wrong'?

Let us return to the time when the Universe was first composed of 75% hydrogen nuclei and 25% helium nuclei. This was just five minutes after the Big Bang. The other constituents of the Universe were electrons and photons. These were all in such a constant state of interaction, sharing energy between them, that they behaved like a very high-temperature gas. In general, the photons collided with the electrons, pumping up their energy, although interactions with nuclei, were also frequent. The electrons then interacted with the nuclei and thermal equilibrium was just about achieved.

After the expansion of space had continued for a year, the density of matter in the Universe had dropped and the photons and electrons found it harder and harder to interact with the nuclei. The nuclei began to depart from the thermal equilibrium and behave in a different way. That different way was defined by the influence of gravity.

The atomic nuclei could now respond to the pull of gravity from their neighbours and, gradually, they clumped together. Photon collisions were not out of the picture altogether; they could still exert an influence. For example, the radiation content of the Universe was spread uniformly throughout. Any region in which matter began to clump would obviously become denser. That density would increase the likelihood of photon interactions with the nuclei and this would resist further clumping.

This is the situation in which the Universe languished for 300,000 years. Clumps of matter were growing on all scales – some very large, others relatively small, yet none able to grow too much. At 300,000 years a fundamental change in the Universe took place. This was the first time in cosmic history that neutral atoms could exist in a stable condition.

A neutral atom is one in which the number of protons in the nucleus is exactly equalled by the number of electrons in orbit around that nucleus. In the case of a hydrogen atom the number of electrons needed is one (remember a hydrogen nucleus is simply an individual proton). A helium atom requires two electrons, because its nucleus is made of two protons (with two neutrons in there for good measure).

Because electrons are in orbit around the nucleus and not tightly bound up inside it, they are comparatively easy to knock off. This ionises the atom. Before 300,000 years, the energy carried on the photons that pervaded the Universe were capable of immediately ionising any atom that formed. After 300,000 years, the redshift had taken its toll and the average photon was too weak to ionise an atom. Hence, once electrons had been captured by atomic nuclei they could no longer be removed. Now that the electrons – which had previously zipped around space making a nuisance of themselves by colliding with photons – had been removed, there was little to impede the photons' progress. They could race through space for incredibly long distances before colliding with anything.

Think of this as being the difference between a foggy day and a clear day. On a foggy day you can see only a relatively short distance in front of you. This is because water vapour in the air has reached such a density that light cannot pass through it without colliding with one of those water molecules and being scattered away from its original direction. On a clear day, water vapour is still present in the atmosphere but in order to notice it, you would have to look at something very far away. If your line of sight can encompass several successively distant hills, the furthest hills will look as if they have a very faint net curtain in front of them, turning them a little white and fuzzy. This effect is exactly the same as happens in fog but on a clear day the water vapour content of the atmosphere is so low that you have to look through miles and miles of atmosphere to notice it.

Physicists distinguish between these two types of cases by calling the atmosphere on a foggy day 'optically thick' and the atmosphere on a clear day 'optically thin'. Think of it like this: optically thin means that the gaps between scattering particles are large enough for most of the radiation to slip between the spaces. Some photons of radiation might be scattered by collisions but they are just the unlucky ones. The average distance a photon can travel, before a collision becomes likely, is known as the mean free path.

Any situation in which light or other forms of electromagnetic radiation must proceed through a medium containing scattering particles – be they dust grains, molecules or subatomic particles – can be defined as being optically thick or thin. The early Universe, for example, before the 300,000-year watershed – that cosmologists grandly call the decoupling of matter and energy – was optically thick. After that, the corralling of the electrons by the atomic nuclei made it optically thin. The photons could travel distances so large that their mean free path was actually greater than the radius of the expanding Universe at that time. Since the expansion of the Universe served to decrease the density of matter still further, as time went by, the mean free path became even larger and the photons could expect fewer and fewer collisions. So does this mean that we can find this left-over fossil radiation today? Absolutely we can. It is called the cosmic microwave background radiation and was discovered by Arno Penzias and Robert Wilson in 1964. These two radio engineers were preparing a radio telescope at Bell Laboratories in New Jersey, in order to dabble in the developing science of radio astronomy. Whilst checking the system, they found a stubborn signal that could not be removed, no matter what they did to the equipment. The constant hiss was always there, regardless of where they pointed the telescope in space, so it had to be noise. It could not possibly be celestial in origin. Or could it?

Sixteen years earlier, a small group of theoretical astronomers, led by George Gamow, had used the previously mentioned perfect gas law to predict that, if the Universe had begun in a Big Bang, then, because everything would have been squeezed much tighter together, collisions between particles would be much more frequent and so the temperature of the Universe would have been much higher. So, if the Universe began in a Big Bang, it was a *hot* Big Bang. The fact that it was hot meant that there should have been a lot of radiation produced. They reasoned that the expansion of space would have redshifted this radiation and that, today, it would be a shadow of its former self, having the bulk of its energy carried on electromagnetic waves in the microwave region of the spectrum. If any were to strike molecules in space, those molecules would be heated – not to thousands of millions of degrees, as in the glory days of the very early Universe but to a meagre −270° C.

The microwaves carrying this relic energy could potentially be detected as a faint hiss by radio telescopes. At the same time that Penzias and Wilson were grappling with their 'noise', less than an hour's drive from the Bell Laboratories a small team of astronomers was building a telescope on the roof of the geology building at Princeton University. Their intention was to detect precisely the same thing that the New Jersey duo were so diligently trying to remove. Penzias and Wilson had grown so frustrated at the persistence of the signal that they even considered it was being produced by pigeon droppings in the equipment. The absurdity of the situation was later summed up perfectly by Penzias, who was quoted as saying, 'We have either discovered the Big Bang or a pile of pigeon shit.' Thankfully, it was the former.

So, the cosmic microwave background radiation was discovered serendipitously and this overwhelming piece of evidence for the Big Bang cemented its place in modern astronomical thinking. Hence, we are left with a Universe containing matter in the form of hydrogen and helium and radiation. The matter will gradually clump together because of gravity's irresistible attraction,

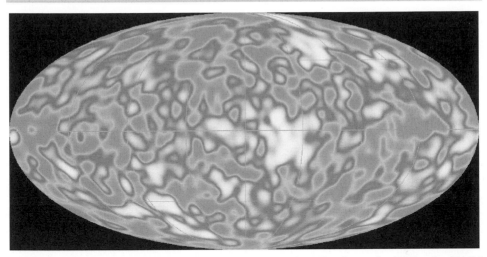

Ripples in the cosmic microwave background radiation. In the early 1990s the NASA spaceprobe COBE (Cosmic Background Explorer) measured the background radiation with unprecedented precision. It found the faint fingerprints left by the clumps of matter that were gradually forming at the time of the decoupling of matter and energy. Had the matter content of the Universe been completely uniform, this image would have appeared as a single uniform shade of grey. These 'ripples', as they were called, are the seeds of the structure we see around us in the Universe today. (Photograph reproduced courtesy of COBE Science Working Group, NASA, GSFC, NSSDC.)

and the radiation will simply travel in space forever, becoming weaker and weaker as the Universe expands (see the illustration above).

From that state of affairs to a Universe full of stars, planets and (in at least one instance) life, there seems to be an incredible gulf. Indeed, there *is* an incredible gulf. It is one that would have been completely impossible to bridge had it not been feasible to side-step the inexorable grip of the second law of infernal dyn ... (sorry) thermodynamics for at least a little while. Let the thermodynamic wars truly begin ...

Chapter Three

Celestial power stations

J ust after the monumental events of the decoupling of matter and energy, the Universe was a place where large, cold clouds of matter were pulling each other together under the force of their own gravity. The chain of events that led to the Sun, the Earth and to you and me, now began in earnest.

The collapsing clouds of gas would become galaxies. In the simplest definition, a galaxy is a cloud of gas that is isolated from other, similar clouds in space. What happens within that gas is largely immaterial to the surrounding Universe. In fact, within each galaxy, gravity continues to perform its magic and the cloud fragments into successively smaller and smaller clouds of contracting gas. At the bottom end of this fragmentation process are the clouds that will become stars. In a typical galaxy like our own Milky Way, there are a few hundred billion stars. The stars inside a galaxy conspire to change the chemical composition of the galaxy, as we shall discuss in this chapter.

THE COSMIC DARK AGES

The Universe, as a whole, was also becoming dark just after decoupling. For the first time in cosmic history, space actually turned black. This seems a strange thing to say. Surely space is dark. Well, think of it this way: space is only dark because there is an absence of light in it. If we could fill it with electromagnetic radiation of visible wavelengths, then it would appear bright to us. Before the decoupling, there was quite a lot of radiation flying around which was in the visible region of the electromagnetic spectrum. Let me explain why.

Hot objects release radiation. This is one of the ways of giving out heat into their surroundings (there's that second law again). In the eyes of a physicist – that is, seen through not just rose-tinted glasses but through *perfectly* rose-tinted glasses, energy is given out in a very characteristic way. This is known as black-body radiation. If one could measure the intensity of radiation given out at all wavelengths and plot it on a graph, it would produce a curve, known as a Planck curve. The wavelength at which the emission peaks can be used to calculate the temperature of the body that is emitting the radiation. For example, the Sun radiates the majority of its energy in the visible region of the spectrum at a wavelength of 550 nm. This means that the visible surface of the Sun is gas at a temperature of roughly 5,800 K. Indeed, any object heated to this temperature will give off radiation with a peak wavelength of 550 nm.

Many factors can mitigate to skew an object's emission away from a black-body curve but they are of little consequence here because the Universe, before decoupling, generated a perfect black-body spectrum. So the wavelength of peak radiation can be used to calculate the minimum temperature that objects in space can possess. At the point of decoupling, the temperature of space was about 3,000 K. This means that the majority of photons were carrying the energy equivalent of electromagnetic radiation with a wavelength of 966 nm. In fact, this is beyond the visible (with its longest, red end of around 700 nm) and into the infrared. Even so, quite a lot of radiation will be carried at red wavelengths of light.

Imagine climbing into a time machine wearing your Acme Big Bang Survival Suit. You are deposited just before the onset of decoupling. The Universe is nothing but a deep red glow all around you, because the red photons are being scattered all over the place by the errant electrons. It is also hot. Accelerating time forwards, you pass through the decoupling. The red glow begins to fade quite rapidly. Not only are there no more collisions taking place to deflect photons into your line of sight but those that you can still see are being redshifted to longer and longer wavelengths. The red glows fades and dies. Space is black. There are no stars here yet – just slowly contracting gas clouds. It is also getting cold. With the redshifting of the radiation, the temperature of space is falling.

If you could 'tune' your eyes to see longer and longer wavelengths, you could have tracked the weakening of the background radiation. The COBE satellite was tuned to 'see' microwaves (wavelength of about a millimetre), so as far as it was concerned, the Universe was not a dark place with the occasional star to break the monotony. Instead, to COBE the Universe blazed with microwaves from all directions (look again at the illustration on p. 27).

To you however, the Universe is looking pretty bleak. There is nothing to see, because nothing is emitting light. These are the cosmic dark ages – an astronomical wilderness that persisted for hundreds of, or even thousands of millions of years. It is a desert that offers no possibility of sustaining human life or anything like it because the only chemical elements around in abundance are hydrogen and helium. As a brief digression, however, let us return to the *Star Trek* cliché in Chapter 1 and ask the question: 'What about life, but not as we know it?'

We are familiar with just one example of life: that which is present here, on planet Earth. However, if we are not careful, this can make us hopelessly chauvinistic. It is very dangerous to draw conclusions from a single statistic. For this reason we must avoid saying that, since life on Earth is based on the chemical element carbon, life must always be based on carbon. I will confront this possibility head-on in Chapter 6, when I discuss the prerequisites for life.

Despite my wanting to keep an open mind, even *I* admit that life based only on the chemicals hydrogen and helium is a difficult thing to imagine. The humble hydrogen atom can bond to only one other chemical and all it would have to choose from is itself or helium. To return to the use of teenagers on Saturday mornings for an analogy: helium is rather like the average adolescent in those twilight hours before midday – inert. You have a hard time persuading it to do anything. So, no real hope there for anything as exciting as life. In the last chapter I described how the Universe could make helium out of hydrogen for a few minutes after the Big Bang. During this time, very small amounts of one or two other elements were also made – most notably, the element lithium – but the amounts were simply too small to have any effect.

The seminal astronomer and writer, Sir Fred Hoyle, imagined, in his wonderful 1950s novel *The Black Cloud*, a living being that was essentially a gaseous cloud. Even in his description of this fictional entity, however, Hoyle was forced to rely on heavy elements to make the being plausible.

Perhaps the best reason for supposing that life-forms cannot evolve from the primordial chemical abundance of the Universe is that, if they could, they would now populate the Universe in vast numbers and be very obvious. Remember that the Principle of Plenitude (Chapter 1) stated: 'Anything that *can* happen *will* happen'. If sentient life-forms could form from hydrogen and helium floating in space, I am certain we would have found them by now. Always one to hedge my bets, however strongly I have just made an assertion, I do acknowledge that, if such life requires unusual conditions (such as David Brin's creatures who live in the chromosphere of the Sun in his superb novel, *Sundiver*) then it is conceivable that we have yet to find them.

For the moment, however, astronomers have made a great study of clouds in space. None of the clouds behave in ways which would suggest anything other than the blind forces of nature at work. The clouds simply collapse together and form stars – as I will eventually describe in this chapter, once I have ceased to digress. I will also try to avoid any more *Star Trek* catch-phrases (at least, for the rest of this chapter).

Having watched the Universe turn dark, it is time for you to climb back into the time machine and go home. Set the co-ordinates ... Final checks ... Engage! (oh dear – just blown the *Star Trek* promise).

Exactly when the dark ages ended and the celestial renaissance began is still something for debate. The continued observation of the Universe will eventually yield this answer. In the meantime, it seems fair to say that one billion years after the Big Bang, the wide-scale process of star formation was in full swing and that it has driven the evolution of the Universe ever since.

OUR SAVIOURS, THE STARS

Our experience of life – and by that I mean the very fact we live on the surface of a planet, bathed in the light from a nearby star – suggests that stars may be the universal life-givers. They are the celestial power stations that give out the energy needed for life to exist. This is well known here on the Earth. The vast majority of life on Earth derives the energy it needs from the Sun. We are taught in school about the way plants make energy from sunlight, that herbivores eat plants for energy and that carnivores then eat herbivores. A recent and exciting discovery shows that, although this is the overwhelming way of things on Earth, it is not the only way. I have already mentioned (in Chapter 1) the discovery of life-forms that live deep underground. These microbes derive their energy from geothermal sources, instead of from the Sun.

For now, however, let us concentrate on stars. Their links to life on Earth are both gross and subtle. A gross link is the supply of energy described in the preceding paragraph. A subtle link is the fact that stars are essential for the production of Earth's life-giving elements, such as carbon, oxygen and nitrogen. Indeed, stars are the producers of the chemical elements that are vital for the Earth itself to exist.

In order to fully understand how amazing stars are, we will first describe the way in which they appear to flout the second law of thermodynamics. They do this by their very act of formation. In resolving the paradox we will be driven towards a more sophisticated understanding of energy and thermodynamics. Remember, the second law states that a hotter object cannot take energy from a colder one. Think of a cold cloud of gas in space. As it collapses under the force of its own gravity, it becomes denser and heats up. Eventually it achieves such a high temperature and density that nuclear fusion begins to take place in its core, *et voila*! – a star is born. This is observational fact. The problem with this picture begins when you try to understand it using the second law of thermodynamics because immediately you run into questions such as: where does the star's heat come from? ... does it take it from the surrounding space?

Well, if that were to be what happens, it would buy you one sixth of a ticket for eating at Douglas Adams' Restaurant at the End of the Universe (or the Queen of Heart's breakfast table – depending upon your literary upbringing). In other words, it is impossible. The temperature of space, as we discussed earlier in this chapter, is governed by the cosmic microwave background radiation and these days it is so feeble that all it can do is heat molecules to a bitter –270°C. Since its release, the expanding Universe has doubled in size 1,000 times and has spread the energy of the background radiation 1,000 times thinner throughout space. Any collapsing cloud is usually hotter than this to start with. So, just how does a star heat up? And where does the extra energy come from?

The key to explaining the phenomenon of star-formation is to develop a slightly more sophisticated understanding of energy. Energy is not just related to the temperature of an object; it can also be affected by its position in relation to another object. Energy can be tied up in movement, too.

Imagine holding a weight out of an upstairs window and waiting for an unsuspecting passer-by (or not, depending upon your motivations for performing this experiment). When you let the object go, it will fall. The movement it suffers on its way down is an expression of energy but where does this energy come from? Presumably, in order to drop it out of the window, you had to walk it up the stairs. By moving the object higher from the surface of the Earth, you stored the potential for it to fall. So physicists call this property 'potential energy'. You know from common sense that you have to work harder to carry heavy items upstairs than to simply walk up them yourself. So the potential energy which became movement energy was taken from the energy you expended to move the weight up the stairs.

The force of gravity is what causes the weight to fall, so gravity can be seen as unlocking the potential energy of the weight. Movement energy is known as 'kinetic energy'. Imagine what would happen if the weight struck the floor (the other, messier, possibility involving the passer-by will be contemplated in the next chapter). When the weight hit the floor, there would probably be a loud noise. So some of the movement would be converted into sound energy. The movement energy can also be transferred to the Earth itself. Because the Earth is so large, however, it will move only by an imperceptible amount. Anyone who watches or practices snooker implicitly understands the exchange of kinetic energy between colliding objects. The energy of movement can also be transferred into the Earth as heat energy. Think of how friction causes our hands to warm up when we rub them together on a cold, frosty day. This is an excellent example of how kinetic energy can be transformed into heat energy.

So here at last we have a way for collapsing gas clouds to heat up. Before the decoupling, the energy of the photon collisions with particles kept them all spread out throughout space (the cosmic equivalent of walking up the stairs). After the decoupling, when the photon support was removed, the particles could respond to the effects of each other's gravitational fields and move towards each other, turning the gravitational potential energy into kinetic energy. This movement eventually resulted in the collision of the particles and the conversion of their kinetic energy into thermal energy and so the cloud began to heat up.

In fact, there is no mystery in a forming star; just the conversion of energy types – an application of the first law of thermodynamics (energy cannot be created or destroyed). Then, as the young star heats up, it radiates energy away into space in accordance with the second law of thermodynamics (heat flows from a hot object to a cold object).

Having shown that stars are not contradicting the laws of physics (which is lucky, considering how many stars there are around) let us now explore the subtle link between the stars and life on Earth. This is the fact that stars create the chemical elements necessary for all life on Earth.

STELLAR ALCHEMY

In the previous chapter I told you that a chemical element is defined by the number of protons it has in its nucleus and that hydrogen, with its nucleus of one single proton, is the simplest chemical element in the Universe. Helium, the second element of nature, possesses two protons and two neutrons in its nucleus. Stars are experts at converting hydrogen into helium; it is their stock in trade. They do it in their fiery hearts, where temperatures are so high that individual protons can be forced to stick together. This is the process called nuclear fusion. It is made possible by the delicate interplay of the forces of nature that take place within stars.

There are just four forces of nature in the Universe. Gravity is an obvious one and we have already looked at its role in forming a star, so let us briefly introduce the other three. Electromagnetism defines the structure of atoms by determining how electrons are bound to atomic nuclei. It also defines the way in which electromagnetic energy interacts with matter; that is, the absorption or emission of photons. The nuclear forces of nature (of which there are two, known as the strong and the weak force) govern the way in which atomic nuclei react with one another. All of these forces take part in the fusion of elements in a star.

At the point at which a gas cloud begins collapsing, gravity has a free hand to do just about anything it wants. Displaying all the imagination of a nudist at a fancy dress party, all it does is squander this opportunity by simply pulling atoms closer together. Then again, that is just about all gravity *can* do. There are very few party tricks with this force, unless of course, you count black holes, which form when gravity pulls things into incredibly dense lumps of matter. Most of the really good stuff that happens in the Universe (such as stars, life and rock music) is pretty much a direct result of the other three forces of nature (okay – maybe not rock music, directly).

If gravity were the only force of nature, these collapsing gas clouds would become smaller and smaller and smaller until, eventually, they turned into black holes. The three other forces of nature make sure this does not happen. In general, squeezing gas into a more and more restricted area actually makes the temperature of the gas increase and, when this happens, the individual atoms in the gas rush around faster and faster. In the centre of a forming star this leads to the protons colliding with each other.

Ordinarily, protons do not like getting too close together. The reason for their stand-offishness is that they are all positively charged and, according to the electromagnetic force, like charges repel whilst unlike charges attract. This is similar to the way in which the north poles of magnets repel each other whilst north and south poles attract. If, however, two protons are close enough to each other (and this is the real trick), one of the other forces of nature takes over. The strong nuclear force will bind the protons together very tightly. This binding process is known as nuclear fusion and it produces the energy which causes stars to shine.

It is not easy to get the protons close enough to each other for the strong nuclear force to work its magic, because the closer they are, the greater the force of electromagnetic repulsion they feel. Think of this force of repulsion as being like a mountain. When the protons are widely separated it is as if they are on level ground and can travel anywhere unhindered. The closer they get, the steeper the gradient becomes until, eventually, when they are very close together, each proton is faced with a metaphorical edifice. Should they be able to climb to the very top, they will be home and dry; completely gripped by the strong nuclear force, electromagnetic energy will have been released and fusion will have taken place. If they do not have enough energy to climb to the top, they will simply roll back down the sides of the force 'mountain' to the flat 'ground' beneath.

In order to become the Edmund Hillary and Tenzing Norgay of the subatomic world, a pair of protons needs to be moving very fast indeed so that their energy can carry them up and over the edifice. Remember that, in reality, the edifice is a repulsive force which becomes stronger and stronger the closer the two protons approach each other. They need to have enough kinetic energy to resist this repulsion until they approach within a distance of just one fermimetre (one millionth billionth of a metre). If they can do this, the strong force binds them together.

The kinetic energy of the individual protons is linked to the temperature of the cloud. As this increases, so too does the average speed of the protons. At a temperature of around 10 million K, protons can collide with so much energy that they fuse with one another. After several subsequent collisions, the atomic nucleus of helium is formed with its two protons and two neutrons. The neutrons, by the way, have been formed by the weak nuclear force which converts protons into neutrons and *vice versa*.

The core of a star, in which nuclear fusion takes place, accounts for about 10% of the mass of the star. The other 90% sits on top of the core. It is far from redundant however because, if it were not for this outer layer of gas weighing down on the core, the gas in there would never be squeezed to the temperature and density that is necessary for fusion to take place.

The photons of energy that are emitted by the fusion, push their way out of the centre of the star and through this surrounding envelope of material. As they collide with the surrounding gas, they support it against further gravitational contraction. So, whilst the star is fusing hydrogen, it is prevented from shrinking any further.

There comes a point, however, when the hydrogen in the core of the star begins to run low and the energy being produced decreases. In the core, gravity again seizes hold and the core begins to shrink, driving the temperature up. If the star is large enough (and all of them are, except the tiniest of red dwarfs) the core temperature will eventually reach 350 million K and helium nuclei can fuse.

The temperature has to be greater because, this time, fusion must overcome the mutual repulsion of four protons (two in each helium nucleus) rather than just two in total (one in each hydrogen nucleus). When two helium nuclei are joined by the strong nuclear force, they create a nucleus of the element beryllium. The nucleus is unstable, however, and not even the strong nuclear force can hold it together. It will rapidly separate into its constituent helium nuclei unless a third helium nucleus collides with the beryllium nucleus within a microscopic fraction of a second of its formation. If this string of interactions takes place, a carbon nucleus is produced. The collision of a fourth helium nucleus produces oxygen.

Again, a star will produce energy quite happily until it runs out of helium in its core. Gravity will then squeeze the core and it is here that the life path of the star will follow one of two different ways. If the star is sufficiently massive (and by that I mean that it contains more than eight times the mass of the Sun) the continual squeezing and ignition of another stage of nuclear fusion will eventually build the elements up to iron.

Iron has an incredibly stable nucleus. Despite the continued contraction of the star's core, it never reaches the temperature for the iron nuclei to fuse together. Instead, the electrons and the weak nuclear force come into play and drive the star into a final phase of explosive evolution.

Throughout this chapter I have discussed the atomic nuclei in the core of the star and have ignored the electrons. Well, they are all in there too, rushing around with high energy providing some support against gravity. In fact, the more the core shrinks, the more support the electrons provide, because the faster they are driven to move. In the iron core of a star, the nuclei are squeezed so tightly together that, eventually, the electrons begin not only to collide with the nuclei but also to penetrate them. Once this happens, the weak nuclear force causes the electron to combine with one of the protons in the iron nucleus and they become a neutron. During this process, a particle known as a neutrino is produced. Neutrinos carry liberated energy away from reactions involving the weak nuclear force, in the same way that photons do for electromagnetic reactions. Unlike a photon, a neutrino hardly interacts with matter and so it escapes from the centre of the star and straight off into space. A typical photon takes about a million years to struggle from the core to the visible surface of a star, from where it can finally head off, unhindered, into space.

At very high densities the fast-moving electrons can be absorbed very rapidly by protons in atomic nuclei. When this happens it is rather like knocking the foundations out from beneath a skyscraper. The support the electrons once provided has been removed and the whole structure collapses. As a star collapses, the iron core suddenly becomes a ball of neutrons. This neutron star, as it is called, contains more mass than the Sun but squeezes it into a tiny celestial object with a diameter of about 20 km. The rest of the star rains down onto this ball, striking its surface and setting up a shock wave that travels outward through the collapsing star, blowing it to pieces. The titanic explosion this causes can be seen, literally, halfway across the Universe (see the illustration on p. 37). For a few days, the exploding supernova produces as much energy as all the other hundred billion stars in the galaxy put together. In the extreme environment this causes, the chemical elements heavier than iron are made. During these explosive events, space is littered with heavy elements. Hence the oft-used saying that 'we are all stardust'. With the possible exception of some of the hydrogen inside your body, every other atom has been at the centre of a star at least once before finding its way into you.

As dynamic as these supernova explosions are, they are not the most efficient way to produce that essential element of life on Earth – carbon. Instead we must think back to the other types of star that are less than eight times as massive as the Sun. It is worth noting here that astronomers are not

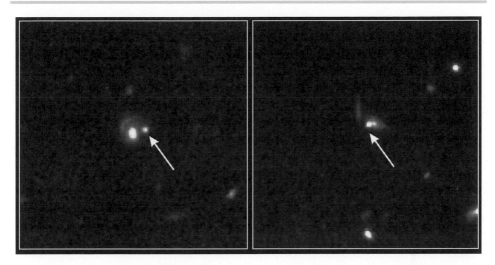

Supernovae in very far distant galaxies. These Hubble Space Telescope pictures show exploding stars in distant galaxies. Approximately one supernova a century explodes in a normal spiral galaxy. When these high-mass stars explode, they are so bright that for a few days they outshine all the rest of the stars in the galaxy. Only the superior vision of the Hubble Space Telescope makes it possible to see the distant galaxies in which these supernovae have erupted. (Images reproduced courtesy of Peter Garnavich, Harvard–Smithsonian Center for Astrophysics, the High z Supernova Search Team, and NASA).

simply practising sidewalk slang when they refer to stars as 'massive'. On Earth we are familiar with the concept of weight. It is a measure of how heavy an object is. Weight is a property of matter that is manifested when an object with mass is placed in a gravitational field. The same object, with the same mass, taken into deep space, well away from any gravitational fields, will become weightless. Hence, all that astronomers are left with is the mass to classify how much matter a star contains.

Low-mass stars do not become supernovae. The reason for this is that they do not contain enough mass to provide the strong gravity needed for them to continue squeezing their core and igniting successive, sustainable nuclear fusion episodes. Instead, what happens is that they enter an unstable phase. During this stage in the star's life, nuclear fusion proceeds sporadically. Internal motion of the gas in the star dredges up material – including quantities of carbon – from deep down near the core. These bursts of energy, known as thermal pulses, drive away the outer layers of the star, scattering the carbon into space.

At the end of this final phase of a low-mass star's life, the outer layers of the star will have been completely thrown off into space, creating something known as a planetary nebula – even though it has nothing whatsoever to do with planets. This extraordinary misnomer came about because William Herschel, who gave them the name, thought they looked like his newly discovered outer planet, Uranus! Considering the frequency with which technical jargon evolves, it is truly unbelievable that no one has renamed these objects.

Both processes have given us today's cosmic abundance of elements. Where once the Universe was composed of 75% hydrogen and 25% helium, now a meagre 2% of the matter in the cosmos is made of elements that are heavier than helium. Just 2%! We are less than the froth on a pint of cosmic beer.

Of this 2% of heavy elements – known to astronomers as 'metals' – the most abundant is oxygen, followed very closely by carbon. Hang on a moment; two of the most abundant elements in the Universe are also two of the most familiar elements for life. Following oxygen and carbon are nitrogen, neon, magnesium, silicon, aluminium, iron, sulphur, calcium, sodium and nickel. I will stop there, because by the time we get to nickel, the abundance is one million times smaller than hydrogen anyway (see the table below).

Having discovered that the majority of elements vital for life on Earth are manufactured in stars, the ultimate question that we seek to answer by the end of this chapter is: from our consideration of stars and the chemicals they seed into space, how plausibly early in the history of the Universe could carbon-based life-forms have appeared? One way to go about answering this question is to try to chart the star formation history of the Universe, because once a star is born, the next thing of importance that it is going to do (within a few million

The cosmic abundance of elements

Element	Relative abundance by number (Silicon=1)
Hydrogen	31,800.000
Helium	2,210.000
Oxygen	22.100
Carbon	11.800
Nitrogen	3.640
Neon	3.440
Magnesium	1.060
Silicon	1.000
Aluminium	0.850
Iron	0.830
Sulphur	0.500
Calcium	0.072
Sodium	0.060
Nickel	0.048

and a few billion years) is die and provide space with a new source of heavy elements.

DEEP FIELDS AND THE HISTORY OF STAR FORMATION

Trying to chart the history of star formation is one of the growth areas of astronomy these days. It has been made possible by the Hubble Space Telescope (HST). In orbit around the Earth, the HST has unrestricted views of the Universe. Unlike the telescopes on the Earth, it never has to stop because of day-time; it simply points away from the Sun and carries on. It also has the privilege of being above the atmosphere. It is a strange quirk of cosmic irony that the light from celestial objects travels vast distances through space virtually unhindered and then, as soon as it hits the atmosphere of the Earth, the air molecules blur it, distort it and generally mess it up completely. Until the space telescope, astronomers simply had to make do with this state of affairs. By putting the HST outside the atmosphere, it can collect the pristine photons and use them to record the most spectacular (and scientifically invaluable) images.

The Hubble Space Telescope regularly returns images that greatly enhance our thinking in astronomy and sometimes the results precipitate a *revolution* in our thinking. One such revolution came from a single image known as the Hubble Deep Field (see the illustration on p. 40). During the early to mid-1990s, in the time leading up to the acquisition of this image, astronomers had been using the HST to take deeper and deeper images of the cosmos. 'Deep' is a term used by astronomers when they want to capture faint objects. This often means that the celestial objects themselves are bright but are rendered faint because they exist at such extreme distances from Earth.

Using deep imaging, astronomers have discovered galaxies that are billions of light years away. This means that the light being received from these galaxies has been travelling through space for billions of years, so the astronomers see these galaxies not as they appear today but as they appeared billions of years ago. The deeper they look, just as the Big Bang theory predicts, the more differences they find between galaxies of the past and galaxies of today. One of the biggest differences is the colour of the galaxies. Once these have been corrected for the redshift of light, it becomes obvious that galaxies were distinctly bluer in the past.

Blue stars are massive and live for only a few tens of millions of years. For there to be a lot of blue light in a galaxy, many massive stars must have been born within a few tens of millions of years of one another. So there is an immediate implication from this single observation of the colour of the faraway

The Hubble Deep Field. On this unprecedented image, more than 3,000 previously undiscovered galaxies were found in an area of sky less than $\frac{1}{140}$ the size of the full Moon. Most galaxies on this image are at extreme distances. An analysis of the colour of the starlight coming from each galaxy can give astronomers an excellent idea about the rate at which star formation is taking place in each one. (Image reproduced courtesy of R. Williams and the HDF Team, STScI and NASA.)

galaxies: they must have been forming stars more rapidly in the past than they are today.

The Hubble Deep Field image was conceived as the ultimate probe of the star formation rate of the Universe. For ten days during the Christmas period of 1995, the HST opened its cameras to a single isolated region of the Universe. It peered at a tiny area of the night sky only $\frac{1}{140}$ the area of the full Moon, to discover more than 3,000 previously unknown galaxies. Apart from sheer numbers alone, this discovery was of paramount importance because the analysis of the image showed that the galaxies are located at many different distances.

By the 'blueness' of the individual galaxies on the Hubble Deep Field, the

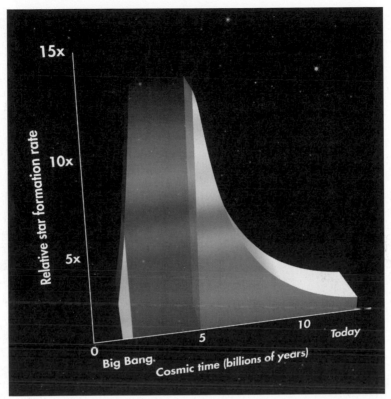

The star forming history of the Universe. Using ground-based observations and Hubble Space Telescope images, the star formation rate of the Universe can be deduced. The peak of activity is observed to take place about 4 billion years after the Big Bang. At this time, 10–15 times more stars were forming than in today's Universe. (Image reproduced courtesy of P. Madau (STScI), J. Gitlin (STScI) and NASA.)

progress of the celestial renaissance can be charted. It suggests a very rapid rise of star formation and that by a billion years after the Big Bang, the star formation rate was probably already twice what it is today. By three billion years, it was between five and ten times what it is today. The peak of the star-formation rate in the Universe probably took place around four billion years after the Big Bang and could have been as high as 15 times that of today (see the illustration above).

From our perspective in cosmic history, we are very much on the Universe's downward slide into obscurity. It seems safe to say that most of the stars that the Universe will ever make, have already been formed – talk about arriving late for the party and missing the fun! From a scientific point of view, a very interesting question is 'How soon could we have arrived on the scene?' Assuming that all other things are equal – that it would take 4.6 billion years from the formation of the Sun and the Earth to the emergence of intelligent humans – at what point were there enough heavy elements in the Universe to permit the formation of terrestrial planets and, subsequently, life?

THE SEEDS OF LIFE

The key to understanding this is to consider the lifetimes of stars. How long does each type of star take to go from formation to death and the seeding of space? As I hinted at earlier, this is entirely dependent upon the mass of the star. The mass determines the temperature of the core of the star. The more mass in the star, the greater the force of gravity squeezing the core and the higher the temperature will soar. The nuclear fusion reactions are very sensitive to temperature. The higher the temperature, the faster the reaction will take place. So, despite the fact that the core of a high-mass star contains more hydrogen fuel than the core of a low-mass star, the more massive star will fuse its way through its hydrogen in a shorter time than the low-mass star and will live a shorter life.

Once a star has begun to fuse hydrogen into helium it is said to be on the main sequence and no matter what the mass of the star, this is where it will spend the majority of its energy-generating life. Through careful observations, astronomers have developed an approximate equation that, for a given mass, can determine the main-sequence lifetime of a star. The equation shows that the high-mass stars (O-type) have lifetimes that are measured in a few million to a few tens of millions of years. Stars like the Sun (G-type) live for about ten billion years and lower-mass red dwarf stars (K- and M-types) live even longer.

It is therefore immediately obvious that low-mass stars like the Sun and the red dwarf stars cannot possibly be responsible for the carbon in the Universe, because they live too long. With the evidence suggesting that the star formation-rate peaked around four billion years after the Big Bang, that leaves, at most, ten billion years between then and now. So the low-mass stars which were formed at that time are only now beginning to end their lives. Clearly there needs to be enough carbon in the Universe, 4–5 billion years ago, when our Solar System formed.

The high-mass stars certainly live short enough lives to have seeded the Universe in time. In fact, high-mass stars are over and done with so quickly that, if it were down to them, space would have been habitable for carbon-based life at the same epoch as the peak of the star-formation activity. The available evidence, however, suggests a slightly different scenario. Between the high- and low- mass stars are the intermediate-mass stars. These are much better at producing carbon than are the high-mass stars, because they cannot fuse it further into heavier elements.

Let us classify intermediate-mass stars as those containing between two and seven times the mass of the Sun. These are not massive enough to create supernovae and so, hopefully, will retain a lot of their carbon content. Then, when the planetary nebula forms, the 'dredging-up' of material, that I talked about earlier, throws the carbon out into space.

There will be a time lag between the peak star-formation activity and the seeding of space based upon the main-sequence lifetime of these intermediate-mass stars. In terms of their classification, they are the A- and F-type stars. The main-sequence lifetime equation shows that their typical lifetime is between one billion and 100 million years. So, in fact, between the peak of star-forming activity in the Universe and the formation of the Solar System, generations of intermediate-mass stars could have lived and died, building up the carbon content of space.

Therefore, the available evidence suggests that life could have been expected to form in the Universe when it was half its present age, somewhere around seven billion years ago. So, it is perfectly conceivable that alien life-forms could have developed on a planet 6–7 billion years ago. With this in mind, I believe that the assumption of temporal mediocrity is on solid ground; the human race exists at no special time.

Having determined that the Universe is a place which has probably been suitable for life to develop for at least 6–7 billion years, it seems sensible to examine what science knows about the origin of life. But hang on a moment; let's not put the cart before the horse. What actually *is* life?

Chapter Four

Life? Don't talk to me

about life . . .

To a physicist, complexity is to be avoided at all costs; it ruins the idealised theories of how matter should act. To a biologist, complexity is just par for the course; something that (pardon the pun) must be lived with. To an artist, complexity is everything because with it come unpredictability, inspiration, creativity, originality and surprises – all those things that make the arts so rewarding.

Life is complex. In fact, life is not just complex – it is *hideously* complex. So you now understand why I, a physicist (albeit an astrophysicist), choose to use the most famous quote from Marvin the paranoid android – creation of writer Douglas Adams, in his book *The Hitch-Hiker's Guide to the Galaxy*. Being an astrophysicist, I am often frustrated by the way the predictions of the perfect laws of physics are distorted by reality. There sometimes seems to be a never ending stream of complicating factors to be taken into account before the straightforward laws of physics can be applied to celestial objects. The forlorn faces of my students often convey this more perfectly than words ever could.

Biologists are much better at accepting the complicating factors and being less fazed by them. This is the art of the 'softer' sciences. Now, do not misunderstand me; that phrase does not mean that they are any easier than the 'hard' sciences like physics and chemistry – just that the predictions made in biology are, by necessity, not so rigidly carved in stone. Instead, think of them as being carved in clay to give a basic outline that is capable of being moulded by complicating factors.

For the moment, however, let me remain a stubborn physicist and wonder whether the essence of life can be distilled into a single sentence. I am going to assume that the complexity of living organisms is clouding our mind a little here and that, somewhere, there is a single indivisible life quality; a single quintessential 'something' that will allow us to quantify life.

Reducing something to its bare essentials is a practise of science known as 'reductionism'. It has served science well since the time of Isaac Newton, who showed that all motion in the Universe could be 'reduced' to a triplet of rules – Newton's three laws of motion. In the context of a definition of life, reductionism may not be the answer, as will soon become apparent. We may never be able to complete the sentence, 'Organisms are alive because of a specific physical property and that property is ...' Instead, we may have to abandon the old ways and appeal to new methods of scientific thinking.

So, before the next chapter's investigation of the way in which life came into being on our planet, let me spend this chapter exploring what life really is.

All of us can recognise life and no one would refute that there is a profound difference between a stone and an animal. In general terms, the stone (an inanimate object) does boring things, whilst the animal (an animate object) does interesting things. Rocks sit around – unmoving, unthinking, unfeeling. Some people do this as well but for the purposes of this book we will neglect such couch potatoes. Returning to the comparison of rocks and animals: rocks do not rifle your picnic basket, nor do they run for cover in the rain. Yet at the atomic level, both the stone and the animal are made of atoms. Even more confusingly, the chemical elements found in rocks will be broadly similar to the chemical elements found in animals, plants and, indeed, human beings. So what is it that changes the non-living into the living?

THE VITAL SPARK OF LIFE

The scientists of antiquity assumed that living matter required some special essence that was absent from all else. These ideas were encapsulated in the scientific doctrine known as 'vitalism'. In the eighteenth century the study of life's chemistry had led to the firm belief that it was beyond the scope of humanity to synthesize organic molecules. Hence, it was believed that some vital force, unknown to humankind, must infuse matter with the spark of life.

In 1786, the assistant of Luigi Galvani, Professor of Anatomy at the University of Bologna, stumbled across a remarkable discovery. Giovanni Aldini was dissecting a frog in a room in which there was an electricity-generating machine and saw the amphibian corpse twitch when he touched it with a scalpel. Aldini told Galvani, who confirmed the reality of the effect. Believing that this might indicate that electricity was the vital spark of life, Galvani turned his laboratory into something that must have resembled one of Stephen King's worst nightmares. Animal corpses, in various stages of dissection, were hooked up to electrical equipment and made to twitch. No wonder, then, that in 1818 Mary Shelley wrote *Frankenstein,* in which the scientist of the title discovered how to use electricity to bring an inanimate being to life. (As it is based on the extrapolation of a scientific notion, *Frankenstein* is probably one of the first pieces of science fiction ever to be written – but that is beside the point.)

Despite an initial flurry of activity, the theory of electricity as the vital force was getting nowhere, even after several decades of scientific study. The story takes its next turn in 1828, when Friedrich Wohler successfully synthesized urea, as he put it, 'without requiring a kidney or an animal, man or dog!' This was the first synthesis of an organic molecule by inorganic means. The important aspect of the feat was that there was no vital force involved – only chemistry. The importance of this discovery was immense and science began to change its viewpoint on life. It began to appear as if the defining quality of life was not a vital force but that, somehow, it was wrapped up in chemistry. As a result, the nineteenth century saw an explosion in the study of organic chemistry which spilled over into the twentieth century and continues to this day.

In the 1870s – at the same time as the science of organic chemistry was booming – an Austrian monk and secondary school teacher was performing world-class research that began the study of genetic inheritance. Unfortunately, his work would have to wait about 30 years before being recognised as such an important step in the field. Gregor Mendel cross-bred sweet pea plants in an attempt to understand how traits such as flower colour and plant size were passed on from generation to generation. His observations led him to propose four laws of inheritance. Central to these laws was the assumption that the information about an inherited trait was carried by a single factor. We now call this single factor a 'gene'. In essence, Mendel's four laws state that

Traits are governed by genes which can exist in one of two forms.

Each organism has two genes for each trait, one inherited from each parent. These two genes may be identical or different.

If the two genes are different from each other, one will dominate the other and cause its trait to show in the organism. The overwhelmed gene is then known as a recessive gene.

If the two genes are recessive, the recessive trait is displayed by the organism.

This work built perfectly on the work of Darwin, whose theory of natural selection was referred to in Chapter 1. Darwin's work in the South Pacific led him to believe that creatures could adapt to their surroundings. The obvious way that such an adaptation could take place would be if successive generations could change their traits. Mendel's work on the inheritance of traits is therefore superb experimental evidence of the mechanism of inheritance. It also raised the question: what exactly carries the inherited information? It has to be an organic chemical of some description but which one?

In the first decades of the twentieth century, two types of molecule had presented themselves as probable candidates. Biologists had recognised long thread-like structures inside living cells. These structures, called chromosomes, only became visible when the cell divided. They were then divided evenly between the two new cells. This observation provided a strong belief that the information-carrying molecule was somehow wrapped up in the chromosome. Analysis showed that chromosomes were rich in organic molecules known as proteins and nucleic acids. The search was on to find the molecule – either a protein or a nucleic acid – which could store information and pass it on.

THE DISCOVERY OF DNA

Many biologists concentrated on the proteins but a few focused on the nucleic acids – especially deoxyribonucleic acid (DNA) – when it was demonstrated that it was the principal constituent of chromosomes.

Studies of the structure of the DNA molecule, using X-rays, began hinting at a helical structure but an adequate theory to explain how the constituents of DNA could build such a shape was elusive. Elusive, that is, until 1953, when British physicist Francis Crick and American biologist James Watson solved the puzzle.

DNA is a polymer, which means that it is essentially a long chain of interconnected molecules. Biologists can split the interconnected molecules up into three basic types: a sugar made of five carbon atoms, a phosphate acid and a nitrogen-rich base. The sugar and the phosphate join together to form the backbone of the DNA polymer. For each sugar and phosphate doublet, a nitrogen-rich base hangs off this backbone, chemically bonded to the sugar molecule. Together, a sugar, a phosphate and a base make up a nucleotide (see the illustration on p. 48).

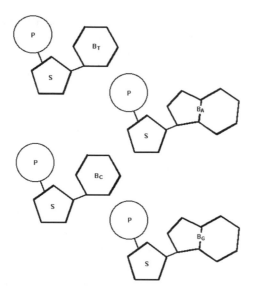

DNA nucleotides. Large numbers of nucleotides join together to form the long-chain DNA molecules. A nucleotide consists of three parts. Firstly, there is a sugar (S). Secondly, there is a phosphate (P). Thirdly, there is a base (B). The base can be one of four types: adenine, guanine, cytosine, thymine (B_A, B_G, B_C, B_T)

Watson and Crick showed that the helical structure of DNA could be represented by two DNA backbones aligned so that the top of one strand (as defined by the lopsided structure of the sugar molecule) was positioned at the bottom of the other strand. The bases that hung from the DNA strands were pointed inward so that they could bond with each other and hold the DNA strands together, pulling them into the now famous double helix.

Next, the real interest presents itself. The nucleotide bases come in four different varieties; adenine, guanine, cytosine, thymine (A, G, C, T). Because of differences in the chemical constituents of these bases, adenine will bond only with thymine. So, everywhere you find adenine on one strand, you always find thymine on the other. Cytosine will bond only with guanine. The concept is known as 'specific pairing' (see the illustration on p. 49). It means that, if you were to pull the DNA strands apart (presumably with a pair of very small tweezers and the strongest magnifying glass you have even seen) and drop one of them in your kitchen sink – which you have previously filled with nucleotides floating around in your washing-up water – as the nucleotides bump into the DNA strand, they will bond only according to the specific pairing rule. Hence they will rebuild exactly the missing DNA strand. If you then drop the other strand into the sink, the same process will happen again. Eventually, where you had one double strand of DNA to start with, you will have two identical

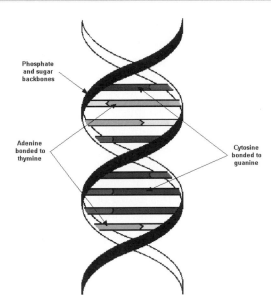

Phosphate
and sugar
backbones

Adenine
bonded to
thymine

Cytosine
bonded to
guanine

The double helix and specific pairing. Two DNA molecules wrap around each other and twist into the famous double helix structure. Holding the two strands together are the bonds between the specific base pairs: adenine and thymine and guanine and cytosine.

ones. The fact that they are identical means that they have copied their information. Watson and Crick realised that their proposed double helix structure for DNA gave it this capacity to self-replicate and copy its information content. In a beautifully understated final sentence in their 1953 paper on this subject, Watson and Crick stated: 'It has not escaped our notice that the specific pairing we have postulated immediately suggests a possible copying mechanism for genetic material.' This would be the equivalent of Neil Armstrong stepping off the ladder of the Eagle lander onto the Moon and saying 'Well, here I am then ...'

So, is DNA life? At the point of its discovery, it was certainly called 'life's secret'. It *is* the secret as far as passing on family traits is concerned but is that, in itself, life?

THE MUSIC OF LIFE

We can answer the above question with an analogy and also highlight how difficult it is to define life. Let us consider music and ask the simple question: 'What is the difference between music and just plain sound?' Sound could be anything: the rustle of the leaves on the trees, a car engine running, or your

DIY-crazy neighbour's electric drill at seven o'clock on a Sunday morning. Music, however, is entirely different. It is an harmonious, melodic or rhythmic sound that means something to us. Yet both sound and music are simply acoustic wave motions that propagate through the air. Music, however, contains information that our brains process into something meaningful. Sound also contains information, just as much as music and frequently more but it means nothing to us apart from identifying its source.

Let us think of the sound as analogous to inanimate objects and the music as the same as living things. The only difference between them is that music and life are put together according to a definite set of rules which produce ordered results. In other words, music is a set of ordered sounds and life is an ordered assembly of non-living parts. You do not have to understand musical composition to recognise and enjoy music. In the same way, although we do not understand life enough to define it, we can still recognise it. Scientists are currently similar to the young upstarts of the pop music business; they know there is something profound in what they are playing but they have yet to get their heads fully around it.

To return to the odd (and probably annoying) habit of quoting rock lyrics in a science book: earlier in his career, Neil Peart (of the *Test for Echo* lyric) also wrote 'There's something here that's as strong as life ...', with reference to music. Oh boy, did he hit the nail right on the head!

There is one major difference between life and music, however, which I shall now address. Having said that both life and music are put together with a set of rules, the question naturally arises, 'Who or what sets the rules?' In terms of music the answer is simple: humans did. There is nothing in nature which makes music; it is a peculiarly human thing. Even birdsong is not strictly music because it is used more like language. But what about life? If we are going to be scientific about this and believe that life began without the intervention of a godhead, then there must be something intrinsic in the laws of nature that allows life to form. In the following chapter on the origin of life we shall look more closely at what this 'something' might be. We also want to know whether it is an incredibly improbable 'something' (which would make life in the Universe rare) or a highly probable 'something' (making life in the Universe common).

For now, let us return to our music analogy. We have not yet been able to say what life is but we can get a little closer to an answer by thinking about CDs that contain music. When music is recorded, the amplitude of the acoustic wave in the air is measured repeatedly, many times every second. The measured amplitude is simply a number, which is then converted into binary form (that is, it is turned into a string of ones and zeros that represent the original number). The ones and zeros are then stored on the CD in the form of microscopic bumps known as pits and smooth areas known as lands. When the CD is played, a laser shines across the disc at a constant frequency and is reflected from the surface. The nature of the reflected light will change, depending on whether the laser strikes a pit or a land and, by detecting these changes, the CD player converts the basic data from a string of ones and zeros into acoustic sound waves via its amplifier and speakers. The same piece of music could also be reproduced by an orchestra working from sheet music, or by the virtuoso musician who simply remembers the music to be played. I am sure that each of us is familiar with the feeling of not being able to get a catchy radio tune out of our head.

So, music can be stored as microscopic pits and lands on a CD, as wave motions in the air, as notes on a stave, or, indeed as electrochemical pulses in the brain. Therefore, although the medium carrying the music can change, it is still unmistakably music. To say music is a CD, an acoustic wave or some ink on a piece of paper is wrong: these items just carry the information necessary to create music.

By analogy, therefore, life is not the medium that contains the information necessary to create it. I made this point in Chapter 1 but it is so important that it is worth reinforcing here. Although I would be the last person to denigrate, in any way, the exceptional work which went into the discovery and understanding of DNA or the work that continues to this day in exploring this amazing chemical construction, it is inescapable that a strand of DNA is not alive; it never has been and it never will be; it is just an organic molecule. Naturally, therefore, we must ask, 'what is the smallest biological structure that is alive?' It is the biological construction known as the 'cell'.

THE LIVING CELL

The cell is the staple building block of an organism. In its simplest form the cell serves as a bag to carry around DNA. Cells such as these are called prokaryotes, which derives from a Greek phrase that means 'before the nucleus'. This is still the form used by today's single-celled bacteria. Within the bag, all sorts of things happen – such as the construction of proteins and eventually the duplication of the DNA and the reproduction of the cell.

A more complex development was the evolution of the nucleus – a bag within a bag. This arrangement allows for much more sophistication in the way the cell builds proteins, among other functions. Eukaryotes, meaning 'true nucleus', are cells which have an inner unit known as the nucleus, in which the DNA resides. In the space between the nucleus and the cell membrane all sorts of other biochemical machinery is found. These discrete units are known as 'organelles'. Different organelles allow the cell to perform different functions. Hence, complementary eukaryotes can work together and form multicellular organisms.

Typically, one of the major jobs undertaken by a cell is the construction of proteins. The proteins are the body's work force and carry out the most amazing variety of tasks. They regulate metabolism; they make up the structure of the body's skin and bones; they make us capable of movement by making up muscles. The protein haemoglobin carries oxygen around in our blood. These are just a few of the more obvious protein functions.

The genes we hear so much about are contained on strands of DNA, called 'chromosomes', which are situated inside the cell's nucleus. Each human cell contains 46 chromosomes. Each gene is the instruction set for a particular protein. So if you imagine that a gene is the equivalent of a sentence, then there must be individual words that make up that sentence.

The 'words' are called 'codons'. Each codon consists of three successive nucleotides on the DNA strand. In the previous section we discussed how DNA is made up of a long chain of nucleotides, each of which consists of a sugar, a phosphate and a base (look again at the illustration on p. 48, for a reminder). The base can be one of four different chemicals (either A, G, C or T).

Therefore, with four nucleotides and three bases in every codon, there are 64 (simply $4 \times 4 \times 4$) different codons based upon the different combinations of bases that are possible. The job of most of the codons in the gene is to specify the type of organic building blocks and the sequence in which they should be put together in order to form that gene's particular protein.

The building blocks are a particular set of organic molecules known as 'amino acids'. They are characterised by having two particular groups of atoms; an amino group consisting of a nitrogen atom and two hydrogen atoms and a carboxylic acid group made up of a carbon atom linked to an oxygen and an oxygen–hydrogen combination. Between these two groups, the amino acid can have chains or rings of carbon which themselves attach to side chains of other atoms. This myriad of possibilities leads to there being hundreds of different amino acids. To make the 10,000 proteins found in a human body, however, we make use of just 20 and put them together in all sorts of different combinations.

If there are 64 possible codons to code for the amino acids and we use just 20, then it becomes obvious that there is redundancy in the code. In fact, several different codons can code for the same amino acid. There are also 'start' and 'stop' codons which are positioned at the beginning and end of each gene. Perhaps most remarkably, there is also 'junk' DNA. These are stretches of nucleotides that can occur between, or even within, genes but which do not code for proteins. Their function remains a mystery as they are apparently meaningless but the startling truth is that so-called 'junk' DNA makes up 85–90% of the DNA contained in eukaryotic cells. So you may think this book is long-winded but in fact all I am doing is applying the DNA principle to my writing.

In order to make a protein from a gene, two processes must take place. The first is 'transcription' and the second is 'translation'. They are only possible because of a molecule related to DNA, called ribonucleic acid (RNA). It is made up of a sugar, a phosphate and a base but is different from DNA in two ways. The first is that the sugar–phosphate backbone contains only one atom of oxygen, not two. The second difference is that the base thymine is replaced by one called uracil. The other three bases remain the same.

Transcription takes place when the DNA double helix is partially unwound by an enzyme. An enzyme is a protein that is the biological equivalent of a catalyst. In other words, it is 'a mover and shaker' that makes reactions happen. The specific sub-group of enzymes used in transcription are known as RNA polymerase enzymes. One will attach itself to the DNA and locate the start codon of a particular protein. It will then pull apart the DNA strands and attract individual nucleotides of RNA that exist in the cell's nucleus. Using the exposed DNA bases as a template, the enzyme will work its way down the strand, building a single chain of RNA that contains the same information as the gene. Approximately 30 RNA nucleotides are added every second and when the enzyme reaches the stop codon it releases the RNA strand. Behind the enzyme, the DNA strands bond together again.

A strand of RNA created in this way is called 'messenger RNA' (mRNA). It will interact with other enzymes in the nucleus that recognise the 'junk' codons and cut them out, splicing the mRNA strand back together again. Eventually the mRNA strand finds its way out of the nucleus and into the cytoplasm (between the nucleus and the cell wall). In here it encounters the ribosomes. These are biochemical 'machines' made of RNA (termed rRNA – ribosomal RNA) and proteins that translate the mRNA into new proteins. Hence, a ribosome is the site of the second stage in protein synthesis: translation.

The translation process relies on a third type of RNA molecule: tRNA (transfer RNA). These molecules bond to amino acids found in the cell's cytoplasm. Each different amino acid is bonded to a different molecule of tRNA. Each tRNA molecule has an obvious triplet of bases on the opposite end from the amino acid which are the three specific bases that will bond to the codon that codes for that particular amino acid. This triplet is known as the 'anticodon'.

When the mRNA bonds to the ribosome, the building of the protein molecule begins. The ribosome attracts tRNA but only when the correct anticodon to the codon held in the ribosome arrives does a chemical bond take place. The ribosome can read two codons at a time. When a second, correct, tRNA arrives, its amino acid bonds to the first amino acid and the first tRNA molecule is kicked out. The second tRNA molecule now moves to the location of the first in the ribosome and pulls the mRNA with it, bringing the next codon into contact with the ribosome. Within short order, the correct tRNA molecule bonds to that codon – and so the process continues, building the amino acid into a long chain known as a 'polypeptide' chain.

This process goes on and on, building the chain until the stop codon is reached and the polypeptide chain is released. At this stage, it can either fold up to become a protein, or it can interact with other polypeptide chains to become a much more complicated protein.

This is all pretty difficult stuff, especially when you consider that, for the sake of the author's sanity, the previous description has ignored some of the subtleties of the process. A prokaryotic cell – one without a nucleus – is considerably simpler in design because, without a nucleus and other organelles, its ribosomes begin protein synthesis as soon as the messenger RNA has been transcribed from the DNA. It is a less ordered system that is prone to errors and almost certainly represents a simpler stage of life before the more complicated eukaryotic cells evolved.

So the importance of DNA in life on Earth is unparalleled but, in terms of providing a general definition of life, it may prove to be the biochemical equivalent of studying one's own navel. Personally, I would prefer to study other people's navels – but that's another story ...

By the time we arrive at cells they are certainly alive and they display a set of properties. Therefore, life itself may not be reducible beyond this set of properties displayed by matter when it is organised in a certain, special way: into cells – and then into organisms. Individually, some of these properties may be displayed by non-living matter but only matter displaying all of the properties can be said to be 'alive'.

THE PROPERTIES OF LIFE

For a collection of matter to be alive it must display the following four properties: self-preservation, self-regulation, self-reproduction and self-organisation.

Self-preservation is simply a life-form's ability to resist external changes in its surroundings. It can respond to changes and, if they are life-threatening, it can attempt to compensate for them. The self-preservation urge drives animals into hibernation to escape the freezing cold and lack of food, it drives humans not to sit in a sauna that is too hot and it drives the million and one other things that all life does simply to stay alive.

Self-regulation is a life-form's ability to keep itself working at optimum capacity. For example: if humans become too hot, small blood vessels near the surface of the skin expand, allowing excess heat carried in the blood to be radiated away. Also, we perspire and the sweat evaporates, taking excess heat with it. One of my favourite pastimes also falls into the self-regulatory category: eating. All animals are capable of eating in order to provide the body with chemical energy. Our bodies let us know when we need food or drink by triggering signals in our brains which we interpret as hunger or thirst. Imagine if you had to consciously remember to drink or eat. You could be suffering from malnutrition before you even knew about it. Cars are not self-regulatory because they need us to top them up with oil and fill them up with petrol.

In many ways, self-regulatory action is the precursor to self-preservatory action. All living things try to 'keep the machinery oiled' because they then function better. If life's machinery gets too far out of sync., however, death or serious injury can occur. So, if self-regulatory measures do not restore balance, then often more drastic self-preservation measures are required.

Next comes self-reproduction. Reproduction in living things is not like photocopying. A photocopier copies something – most often the information on a piece of paper. A photocopier does not copy the paper itself – just the information. And the photocopier certainly cannot photocopy itself and produce another photocopier. Nor is self-reproduction like the manufacture of a car because the machines making the car are not themselves cars. Only living creatures are able to self-reproduce. In other words, humans make other humans, rabbits make other rabbits and biki-biki beetles make other biki-biki beetles. In the final chapter of this book I will consider the possibilities of robotic spaceprobes reproducing.

As well as the capacity to self-reproduce, living organisms also have the ability to self-organise. This is the ability to take in food and use it as raw material to grow or to replace damaged parts. For example, a compact car with air conditioning does not grow up into a stretch limousine. That is because it is not alive. As another example of what is turning into a rather anti-automobile assault: broken cars do not repair themselves either (oh, that they would ...) because they are not alive and are not capable of self-organisation.

The ability to self-reproduce is wrapped up with the ability to self-organise. In terms of single-celled creatures, the self-organisation of matter leads directly to cell division and hence self-reproduction. In terms of mammals, the mother supplies a location in which self-organisation can take place in a cosseted environment. Eventually the growing life-form can function in the outside world and survive by itself. At this point it is born and self-reproduction has again taken place. In broad terms – the more complex the creature, the more time it needs within its mother, or some other cosseted environment, such as an egg, to self-organise into an independent state.

You may think that your disastrously untidy spouse/child/lodger would not recognise the concept of self-organisation if it were to stand up and spontaneously tidy up its mess whilst whistling the opening fanfare of *Star Wars* but believe me, they are experts at it – and so are you. You (and they) simply do not know it. Self-organisation on the physical/chemical/biological level is something you do not have to think about; it just happens and that makes me suspect that something in the laws of physics must be responsible.

Thinking thermodynamically, something capable of self-organisation is known as an 'open thermodynamic system'. There are two types of thermodynamic system: open and closed. A closed thermodynamic system does not take in or give out energy and/or matter. Remember that in Chapter 2 I stated that the Universe was the ultimate closed system. An open thermodynamic system interacts with its surroundings by exchanging energy and matter.

There are many examples of open thermodynamic systems in nature. Not all of them are alive, however, because they do not embody those other properties of life.

Even a humble saucepan of milk is capable of self-organisation. The saucepan of milk is the thermodynamic system. It becomes an open system when it is placed on a hot cooker so that heat enters the milk from the cooker. As the milk becomes hotter than its surroundings, it begins radiating heat away. So, energy is free to enter and leave the milk. The individual molecules are free to move around in a random fashion. As heat is applied to boil the milk, the molecules move faster and faster. They are moving hither and thither with no preferred direction until the milk begins to boil. Suddenly, the boiling milk begins to roll over and over in a process known as 'convection'. Instead of acting as individual molecules, they begin working together, moving in the same direction to produce these rolling convective cells. It is a spectacular example of self-organisation. But how can they do this? And who tells them to do this? No one tells them; they just do it. It is obviously hard-wired into the laws of physics.

When you pull the plug on a bath, a little whirlpool forms around the plug hole. This too is an example of self-organisation. Water molecules move through the structure continuously and yet the shape of the vortex remains. This is an example of the flow of matter rather than of energy. The same principle is at work in the atmosphere of Jupiter. On this massive scale, a vortex caused by the motion of gas in the atmosphere has persisted for more than 300 years – for most of the time that telescopic observations have been possible. The feature is known as the Great Red Spot and is one of the most famous 'landmarks' in the Solar System (see the illustration on p. 57).

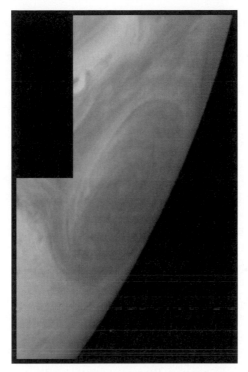

Jupiter's Great Red Spot. This vortex is a spectacular example of self-organisation on a truly colossal scale. The spot itself is about the same size as the planet Earth. (Photograph reproduced courtesy of NASA, JPL, Galileo Project Team.)

Think back to the example of the forming star in the previous chapter. That was also an open thermodynamic system because it was taking in energy in the form of gravitational potential energy as it collapsed, converting that to heat and then radiating it away into space. Later, after nuclear fusion begins, it is chemical energy that provides the star's energy. In the process, the star becomes a more compact, organised structure.

The absolute key to self-organisation is the influx of matter and/or energy. Without it, the structure dies. When the bath water runs out, the whirlpool disappears. When you stop heating the milk, the convection pattern stops.

A life-form is, by necessity, an open thermodynamic system. Think of an organism, taking in food. I do not need to tell you what would happen if you turned that life-form into a closed thermodynamic system by placing it in a sealed box. Self-organisation would stop, the life-form would die and it would begin to decay into a less organised state.

This is, in fact, the big difference between open and closed thermodynamic systems. Open systems can become more organised, whereas closed systems can only become less organised.

ORGANISATION AND ENTROPY

With this concept of organisation in mind, it is time to look again at the second law of thermodynamics (heat cannot flow from a cold body to a hot body) and develop it one stage further to introduce the concept of entropy.

Physicists discuss the change in the organisation of a system using the concept of entropy. A system with a large amount of organisation has a low entropy; that is, entropy is a measure of disorganisation. A system that is totally random, with no organisation, has a high entropy. So if atoms are scattered throughout a large volume of space, as in a giant molecular cloud, that volume of space has a high entropy. When atoms are collected into a small volume of space to form a star, they become more organised and have a lower entropy.

As an open thermodynamic system self-organises itself, so its entropy diminishes. It is a fact of life, however, that any closed thermodynamic system, left to its own devices, degrades and decays. For example; cars require us to maintain them, otherwise they just fall into a state of unusable disrepair. Animate objects, being open systems, continue to repair themselves throughout their lives but after the natural self-organisational processes end (at death, when it ceases to take in energy/matter and so becomes a closed system) the organism becomes increasingly disorganised as it decays.

So, while it is true that certain thermodynamic systems can decrease their entropy, nature, on the whole, likes entropy to increase. In fact, it can be shown that whenever an open thermodynamic system decreases its entropy, the entropy of its surroundings increases. This is an important concept. Although science is incapable of assigning actual numerical values to the entropy of a specific system (and does not really need to, as only changes in entropy are important), in a broad sense, the number of photons in the Universe can be thought of as a measure of its entropy.

To illustrate this point, think once again about the forming star example. Potential energy was transformed into electromagnetic energy and carried away into the Universe on photons. So the entropy of the cloud decreases as it organises itself into a star but the entropy of the surrounding Universe increases as the star radiates the converted energy into space. Increasing the number of photons in the Universe increases the entropy of the Universe.

When an open thermodynamic system and its surroundings are considered, the system becomes a closed thermodynamic system. In such a system, entropy can only increase and never decrease. So, as a human being grows, decreasing its own entropy, the surrounding Universe must pay for this by increasing its entropy by an amount greater than the decrease. This is usually achieved by the production of waste products. In the forming star, the waste product is the photons.

As humans, we are all familiar with certain biological waste products but choose not to talk about them in polite company. This form of waste can be thought of as a low-grade chemical waste. There is also another form of waste product that all of us produce: heat. We are accustomed to car engines and electrical transformers producing waste heat but initially it is strange to think of human beings doing so too – but they do. Heat is a form of low-grade electromagnetic energy because each infrared photon carries less energy than a visible photon or an ultraviolet photon. All waste heat is carried away on infrared photons of energy; so again, the entropy of the Universe increases.

The inexorable decay of the Universe into a state of complete disorder is therefore inevitable and is taking place all around us. Things really do go from bad to worse, despite what the eternal optimists might tell you. To make matters worse, humans are partly responsible because, just by living, we pollute the Universe with waste heat, increasing its entropy and hastening its demise. We can at least share the blame with all the other life-forms on the Earth and with the inanimate open thermodynamic systems that occur in nature.

The completely disordered state is reached at the point when the Universe achieves the thermal equilibrium which I first discussed in Chapter 2. Heat flow will stop, there will be nothing to drive chemical reactions and so no more waste heat will be produced. The entropy of the Universe will have reached a maximum and it will die.

But is this bringing us closer to a definition of life? Yes – I think it is. The process of giving out energy so that a thermodynamic system can decrease its entropy is the key. In school, everyone is taught about the three states of matter: solid, liquid and gas. But science knows of another three more extreme states: two degenerate states of matter that can be created in intense gravitational fields and a high-energy gaseous state known as a 'plasma'. For our purposes, however, we can simply adhere to the usual three. The change of a state of matter to another state is known as a 'phase change'.

Take some water and place it in the freezer. The surroundings are colder than the water, so, in accordance with the second law of thermodynamics, the water gives out energy and freezes into a crystalline solid known as ice. The ice is a very low-entropy object because it is simply a repeated arrangement of atoms, whereas the water was a jumble of individually moving water molecules and so had a high entropy. The properties of the water changed when it underwent a phase change and its entropy decreased as it assumed a more organised structure.

So can it be that life itself is a kind of phase change of matter? Is it simply a new state of matter that manifests the properties of self-organisation, self-regulation, self-preservation and self-reproduction? If this is the case, then life simply emerges as a surprise and cannot be predicted by a knowledge of its individual components or reduced into a sentence.

The only way to test this hypothesis is to put together the correct chemical mix in the correct physical conditions and watch life emerge from the chemical soup. So, ultimately, the question 'what is life?' is intimately wrapped up in the question of 'how did life begin?' and we are ready for the next chapter ...

Chapter Five

The origin of life

To know the way in which life began on this planet is tremendously important in the search for extraterrestrial life. The previous chapter, I hope, presented a flavour of the complexity of even basic life. If life began in a Darwinian way, with simple chance interactions, then, as I explained in Chapter 1, we are very lucky to be alive. If just one of those steps had been different, then everything would have veered off in a totally new direction. So to assume that life on this planet is the product of a string of random chemical interactions is to virtually rule out the possibility of life elsewhere in the Universe.

Some biologists believe strongly – in fact, passionately – that this is the case. In their view, just one different chemical interaction on the route to producing the first cell would have doomed it. Life, they say, is a frozen accident; chance caught on the wing.

The actual probability of life developing can therefore be estimated statistically simply by multiplying all those tiny probabilities together. In statistics, the number 1 represents absolute certainty and the number 0 total impossibility. So the probability of the right interaction taking place is always a number between 1 and 0. When you multiply numbers less than 1 together, the answer is always smaller than those you started with. For example, 0.1 multiplied by 0.1 is 0.01.

CHANCE CAUGHT ON THE WING?

Biologist A.G. Cairns-Smith estimates that there are at least 140 discrete chemical steps when building a useful nucleotide sequence such as DNA. He compares getting each step right as the equivalent of throwing a six on a dice. So to build DNA by random chemical interactions is the chance equivalent of throwing 140 sixes in a row. The chance of throwing a six on a dice is one sixth, so the chance of throwing 140 sixes is one sixth multiplied by itself 140 times. It is an incredibly small number. Most pocket calculators will fail to work it out and will give the answer as 0 (or 'E'). Well, we know it is not zero because we are all here pondering the problem. In fact, it is a decimal point followed by 109 zeros and then a one. Scientists write this figure as 1×10^{-109}. It means that, on average, for every 1×10^{109} chemical reactions that occur in the Universe, one DNA strand will be built. Do not forget, this gets us only as far as DNA, not to the far more complicated living cell.

'But,' you cry, 'the age and size of the Universe makes even the tiniest probability a virtual certainty!' Let us look at this idea more closely. In Chapter 2 I stated that the Universe is about 15 billion years old. That means that about 5×10^{17} (5 followed by 17 zeros) seconds have lapsed since the Big Bang. This result is derived from an eight-year study of the cosmos undertaken with the Hubble Space Telescope. In the last chapter I stated that about 30 chemical reactions per second are possible inside living cells. As a rough estimate, there are 1×10^{80} particles in the Universe. These make up everything – the stars, the planets, the interstellar dust and gas, you and me. Let us say that each of these 1×10^{80} particles can partake in 30 reactions per second. Then the number of chemical reactions that have taken place since the beginning of the Universe is $5 \times 10^{17} \times 30 \times 1 \times 10^{80} = 1.5 \times 10^{99}$.

Since the probability of building DNA by chance requires an average of 1×10^{109} reactions and there have only been 1.5×10^{99} since the Universe began, there is an obvious shortfall. Remember, the probability of the subsequent chemical reactions leading to a living cell have yet to be included in the calculation. So if Darwinian evolution and contingency stretches all the way back to the origin of life, the probability of life in the Universe is so small that we should not expect to find it anywhere else. Life in one location is unlikely enough but in two or more locations it is inconceivable. We are indeed a frozen accident of such low probability that it boggles the mind. It is no wonder that Sir Fred Hoyle once drew the analogy of the chance appearance of life on Earth with the concept of an aircraft hangar full of the dismantled pieces of an aircraft. Imagine our incredulity, his reasoning went, if a tornado swept through that hangar and, in its wake, left a perfectly assembled 747 airliner.

Could we really be the only life in the Universe? Well, before we start dismantling the SETI receivers in despair, there are other views about how life may have begun. As much as I respect the biologists, I cannot believe that life is a fantastically unlikely chance caught on the wing. Once life-forms are around, then evolution is indeed your guide – but not before. It renders the origin of life too improbable for my taste. It makes buying lottery tickets look like a wise investment. Thankfully, there are a growing number of biologists who are also beginning to question this received wisdom.

So, what exactly are the alternatives? Well, I am afraid I cannot possibly advocate the obvious religious one. I was unhappy saying that life's origin was an incredibly-low-probability event, so I am hardly going to be happy saying that it was an impossibility that needed a miraculous guiding hand to instigate it. Instead, I shall concentrate on a way of increasing the probability of the origin of life.

Therefore, for the duration of this chapter I am going to advocate a viewpoint that life itself is something known as an 'emergent phenomenon'. I mentioned this concept at the end of the previous chapter but now it is time to confront it head-on. It is a comparatively new way of thinking and it challenges the old view of reductionism. An emergent phenomenon is one that cannot be predicted from a knowledge of its components. Indeed, it may be impossible to predict it; it just happens when a sufficiently complicated system of components begins to work together. You can reduce systems as much as you like but anything you still cannot explain must be emergence. Let me be specific. Although none of the atoms which make up your body are alive, you most certainly are alive – or doing a very passable imitation in order to be reading this book.

Other examples of emergent behaviour include the wetness of a liquid. The individual molecules of the substance do not exhibit the quality but a collection of them does so. The behaviour of ants in a colony is another good example. Individually, an ant is an insignificant insect but when it gets together with a few thousand of its buddies, the resultant ant colony displays a remarkable, emergent, 'group intelligence'. The behaviour of societies and packs of dogs are other examples. Humans have intuitively understood the concept of emergent behaviour for a long time because it gives rise to the saying that 'the whole is greater than the sum of its parts'. In the case of rock group Rush, each of the three musicians are accomplished virtuosos in their own right but when they take to the stage together ... wow! Their combined talent conjures a special, spine-tingling magic from the ether. I am sure each of us feels that way about certain groups of people – be they musicians, actors, writers, or even scientists.

Somewhere between the simple, inanimate laws of physics and chemistry and the complex behaviour of biological systems, life emerges. It therefore seems to me that the key to understanding the origin of life is interdisciplinary science, with emergent behaviour bridging the gaps between physics, chemistry, geology, astronomy and biology. All have their role to play in answering what must be one of science's most fundamental questions: how did life form on this planet?

CONDITIONS FOR EMERGENCE

I have already mentioned Newton's laws several times. Their publication in 1687 was a landmark in science. I have stated how they reduced motion to a set of logical and mathematical rules and also the way in which they gave rise to the concept of the clockwork Universe, in which everything was predictable. Since the middle of the twentieth century, however, scientists have come to recognise that not everything is as tidy and predictable as might have been believed.

Chaos theory studies phenomena which cannot be accurately predicted. In a Newtonian system – sometimes called a 'linear' system – a slight deviation of the input leads to a slight deviation of the output. So, approximate conditions can produce approximately the right results. A chaotic or non-linear system is not like this. A slight variation of the input can have dramatic effects on the output. Chaos is by no means a random, disordered state; it is simply one that rapidly diverges from predictions. Take that quintessential British preoccupation: the weather. It is a non-linear system and so is fundamentally difficult to predict. With all the knowledge gleaned from satellites and ground stations, a meteorologist can still only predict the conditions for a day or two ahead at the most. Anything else is little more than educated guesswork, as those of us, caught without an umbrella at the ready, know only too well.

Between completely linear and totally non-linear systems we find dissipative structures. These are basically open thermodynamic systems which come to order by the process of self-organisation. The self-organisation (discussed in the previous chapter), which is so important to life, takes place at the edge of chaos in a regime that is delicately balanced between boring predictability and unruly chaos. With self-organisation comes the possibility of emergence.

As a dissipative structure forms, it manifests wholly new properties that were not present in the constituent material. The amazing thing is that, at the edge of chaos, a dissipative structure can repeat the same process of self-organisation again and again. Take the example of water spiralling down a plug hole. If you swish your hand in the water and disrupt the mini-vortex, it forms again when you remove your hand. This compulsion to self-organise is explained by the esoteric concept known as an 'attractor'. Attractors are an expression of the non-linear characteristics that drive the dissipative structure to repeat the organisation over and over again. When a pattern is repeated in a system we say that the system is in the 'grip' of an attractor. The more complicated the non-linear system, the more complicated the attractor until chaos sets in, in which case we say that the system is now in the grip of a 'strange' attractor.

Since life is a biochemical dissipative structure we must face up to the fact that it, too, may be in the grip of an attractor. A sufficiently complicated chemical system, under the right conditions, will be driven to create life with all its fabulous emergent properties intact. More than this; if the conditions exist somewhere else, life will be created there too. There is no chance involved. Life is an inescapable consequence of nature. I firmly believe this. In fact, I believe it so strongly that I am going to devote the next paragraph to this statement alone.

Life is an inescapable consequence of nature.

In Chapter 1, I discussed the Principle of Plenitude (anything that *can* happen *will* happen) and the Copernican Principle (Earth is nothing special). If it can be proven that life is in the grip of an attractor, then the Copernican Principle applies to biology too and by the Principle of Plenitude our Galaxy and all the others will be full of life.

What are the right conditions? Well, a steady supply of energy for one thing. Remember that dissipative structures are balanced on a knife-edge between simplicity and chaos. This remains the case with life today. Too little energy and the organism will not be able to make use of it. Too much and the organism will be harmed by the sudden uncontrollable influx of energy. This is why, at a cellular level, ultraviolet radiation wreaks havoc. It supplies so much energy to a cell that it harms it, causing it to die or to turn cancerous.

Recall the example in Chapter 3, in which we envisaged dropping a weight from an upstairs window onto an unsuspecting passer-by. It hits the head of your victim, delivering a large quantity of kinetic energy – and you do not need me to describe the resulting carnage. The moral of the story is that too much energy, delivered too quickly, can be very damaging for a fragile dissipative structure such as the human body. This is why we take in food and let our bodies extract the chemical energy, rather than plug ourselves into the mains for that little early morning 'pick-me-up'. Some science fiction writers, however, love this idea and play with it by having the square-jawed Captain Testosterone watch in horror as invading aliens from the planet Zog absorb his laser fire and use it to make themselves more powerful. Of course, the aliens prove no match for Captain T. when he throws down his Blastomatic^{TM} laser machine gun and takes them on, 'mano-a-tentacle'. But I'm getting carried away again ...

Another thing to realise is that the present conditions for life on Earth may not be the same as those which were necessary for life to begin. In fact, the evidence suggests that life started in an atmosphere that we would find totally unbreathable and with a menu that would make dog food seem positively appetising.

TRACING LIFE BACK TO ITS ROOTS

Assuming that life began from non-life, we should expect that simpler life came before more complex life. So can we put living forms into some sort of order? Intuitively, simple prokaryotes must have existed before eukaryotes with nuclei, which must have existed before multicellular organisms.

Tracing the evolution and development of multicellular organisms is rendered possible by the existence of the fossil record. As creatures die, their bodies are gradually covered by sediments and the hard parts of their structures, bones and shells, are preserved. By using radioactive techniques to date the rocks in which fossils are found, the picture of evolution on the Earth can be traced. A fascinating picture it is too – as I shall discuss in Chapter 9, when I consider the rise of intelligence on both Earth and other worlds.

Under special circumstances, even single-celled creatures can leave fossil imprints. These are called microfossils. According to the fossil record, single-celled creatures were the first to appear on Earth, with multicellular life emerging billions of years later. Old sedimentary rock called 'chert', in Western Australia, has been found to contain fossils of bacteria that look very similar to bacteria of today. Dating of this rock has shown that it formed (and therefore incorporated the bacteria) 3.5 billion years ago.

In Western Australia and in Swaziland there are are rock structures that resemble present-day stromatolites. These are sedimentary structures built by layers of bacteria living on top of one another in shallow waters. The ancient examples of these structures date from 3.3 to 3.5 billion years age. The best place to see examples of these archaic bacterial colonies is in Shark's Bay, Western Australia.

There is also some chemical evidence in 3.85 billion-year-old rocks that life must have been present, although actual fossils have not been found. It is just about at this point that our luck runs out because there are hardly any rocks older than this on the Earth. The Earth is in a constant state of volcanic renewal, with older rock being directed back down into the molten interior for geological recycling. Sadly, this means that the fossil evidence of life's emergence on Earth is lost forever.

Nevertheless, I have endeavoured to explain why scientists believe that life in cellular form was present on Earth 3.5 billion years ago. This is estimated to be 1 billion years after the main bulk of the Earth formed. If I press on with the story and tell you how the Earth formed, it will become apparent that life's origin is probably confined to a window of a few hundred million years, starting some time around 4 billion years ago.

THE FORMATION OF THE EARTH

Think back to Chapter 3, in which I described the birth of a star as the almost inevitable consequence of the collapse of a gas cloud in space. The gas cloud collapsed under the force of its own gravity, heating up the gas so much that, eventually, nuclear fusion ignited in the core of the collapsing cloud and a star was formed.

There is, however, one complicating factor that I did not mention – simply because I did not need to consider it at the time but, in reality, this complicating factor is the reason why there are planets around stars. It is rotation.

All the gas clouds in the Galaxy rotate. They only rotate very *slowly* but they all rotate. When the cloud begins to collapse, the rotation increases in exactly the same way that an ice-skater can spin faster simply by pulling in his/her arms whilst turning around. The increased velocity of rotation makes the collapsing cloud pancake into a flattened spinning disc. The diameter of such a disc is about 1,000 times the distance between the Earth and the Sun. In astronomers' lingo this is 1,000 AU (astronomical units). The Hubble Space Telescope now studies examples of these structures on a fairly regular basis (see the illustration below).

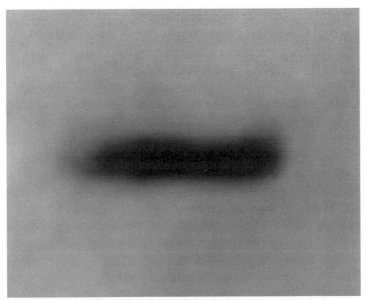

A dust disc around a young star. The dark silhouette in this Hubble Space Telescope picture is a dusty disc seen edge-on. It surrounds a young star, which cannot be seen directly but is illuminating more tenuous dust above and below the disc. In the inner confines of the dust disc, planets are suspected to be forming. (Photograph reproduced courtesy of M. J. McCaughrean (MPIA), C.R. O'Dell (Rice University) and NASA.)

Another complicating factor is that not all of a cosmic gas cloud is, in fact, gas. About 1% of the mass of a gas cloud is composed of dust particles. These dust particles are predominantly made of carbon, silicon and oxygen. They may also be coated with a frozen layer of water, carbon dioxide, methane and ammonia in varying proportions.

Once in the relatively dense environment of a disc, the dust particles can collide and stick together. Through a series of such 'sticking' events, the matter in the 1,000 AU disc 'settles' into a smaller, denser disc of about 100 AU is diameter. At the centre of this disc is the forming star. Gas flows through the disc onto the star, making it bigger and pushing it ever closer to the moment of ignition – when fusion begins in its heart. In the disc itself, the solid particles continue to grow into the size of the present-day asteroids and greater. In the context of planet formation, astronomers call them 'planetesimals'.

When planetesimals collide they do not really 'stick'. Instead, the impacts melt them and they resolidify as a new, bigger rock in space. Eventually, after 50,000 years or so most of the planetesimals will have collided and grown so much that they will be recognisable as planets. For a basic definition of a planet, let us say it is large and travels in a more or less circular orbit around a star – in our case, the Sun. In contrast, a moon is a small celestial object in orbit around a planet.

This is roughly the state that the Earth and the rest of the Solar System found itself in 4.6 billion years ago. The Sun had ignited and all the planets were present in recognisable form. In terms of life, however, it was the following half a billion years of history that were of crucial importance to life's origin. During this time all the chemicals necessary for life on Earth were deposited on our world in a continual bombardment by the remaining planetesimals.

This material rained down on the planets, pock-marking them with craters. Looking at Mercury and the Moon, the craters are still clearly visible today because neither possesses an atmosphere to erode away the rocks. Venus, Earth and Mars possess substantially fewer craters – not because they escaped the bombardment but because the craters have been removed by geological and atmospheric processes.

The planets of the inner Solar System formed almost exclusively from rock and metal because the Sun's heat would not let other material with lower melting points, such as water, condense into solid particles. However, it cannot possibly have escaped anyone's notice that two thirds of the surface of Earth are covered in water. If water could not condense in the inner Solar System, how did our planet come to possess so much of it?

This problem is related to the origin of life. Biologists refer to life's crucial

Comet Hale–Bopp passed through the inner Solar System in 1996. This object has remained largely unchanged since the formation of the Solar System 4.6 billion years ago. When it passes through the inner Solar System, the Sun's radiation sublimes some of the ices away into space. (Photograph reproduced courtesy of Jim Collett, University of Hertfordshire.)

chemical elements as the CHNOPS elements – an acronym for **C**arbon, **H**ydrogen, **N**itrogen, **O**xygen, **P**hosphorus and **S**ulphur. The immediate realisation is that phosphorous and sulphur can be derived from the Earth's rocky material in the abundances we see on Earth today; but where did the abundances of the other four come from?

One of the guaranteed crowd-pullers for any observatory's open night is a good comet in the sky (see the illustration above). These ghostly apparitions with their tails stretching across the sky are beautiful reminders that we live in a dynamic Universe in which things are constantly changing. In some ways, comets are the icy equivalents of asteroids. Both classes of object are the remains of the planet-building phase of Solar System history. The asteroids formed relatively close to the Sun and so are made of rocks and metals. The comets formed in the outer Solar System, far away from the Sun. Containing rocks and metals, they will also have formed from icy material. To an astronomer, ice is not just something to put in a glass of whisky – although that use is acknowledged to be one of the most important. Rather, the astronomical ices are not only water (H_2O) but also methane (CH_3), ammonia (NH_4) and carbon dioxide (CO_2) – all pretty good sources for the two thirds of the CHNOPS elements that cannot come from rocks. So, whereas the use of all astronomical ices in whisky is not recommended, we can see that their use in the origin of life is of crucial importance.

In the heavy bombardment phase that ends planet formation, some of the colliding planetesimals will have originated in the outer Solar System. The collision of these comets with the inner planets is thought to have supplied somewhere between 50% and 100% of the total amount of astronomical ice now found in the inner Solar System. The two smaller worlds – Mercury and the Moon – did not have enough gravitational pull to hang onto these substances once the heat from the Sun had melted them into liquids and gases, so they just evaporated away into space.

On the larger worlds of Venus, Earth and Mars, the incoming ices were retained and helped to form the atmospheres. The rest of the gases forming the atmospheres were released when the rocks making up the planets were melted. This melting can be caused by shock-waves from the impact of the planetary debris, or by the decay of radioactive elements in the planets' rocks.

Once an atmosphere of sufficient density builds up, oceans become possible. This is because when ice melts, each water molecule – by absorbing a photon – is supplied with enough energy to break the chemical bonds with its neighbouring molecules. Whatever energy is left over from breaking these bonds is transformed into movement. If there is nothing to confine the errant molecule (or the rest of its buddies, undergoing the same ordeal) they will exercise their newly-found freedom by streaming off and becoming a gas. A solid's direct transformation into gas is called 'sublimation'. This is what happens on a comet. Sunlight melts the ice on its surface but because a comet has no atmosphere the ice sublimes into a gas and streams away into space, creating the tails that enchant observers from Earth.

If an atmosphere is pressing down on the solid, however, our errant molecules have to exert energy to push their way into the atmosphere. If they have been supplied with just enough energy to break their chemical bonds but not enough to push their way into the atmosphere, then the molecules will be confined in a chemically unbonded mass – a state of matter which we call a 'liquid'. Hence, lakes and seas are created.

So by the end of the late bombardment, four billion years ago, comet collisions would have helped to create the Earth's atmosphere, caused it to have oceans and seeded those oceans with the CHNOPS elements. Next time you look at a comet in the night sky, it is worth remembering that comets made life possible on Earth.

Some astronomers and scientists believe that more than just the chemical building blocks of life fell to Earth. Sir Fred Hoyle and Chandra Wickramasinghe believe that microbial life itself was brought to our planet on comets. This concept is known as 'panspermia' and was first coined in the nineteenth century to describe the passage of microbial life through the Earth's atmosphere. As Louis Pasteur was performing his experiments to show that life did not spontaneously form from inanimate matter but was transported from place to place on the air, so R.E. Richter suggested that life did not form on the Earth but was brought to our planet in meteorites from space. The term 'panspermia' was retained for both concepts. In recent years the acronym BFS – Bugs From Space – has been used but Hoyle and Wickramasinghe have suggested the term 'cosmicrobia'.

Studies are clearly indicating that bacteria can withstand the harsh environment that is present in space. On 20 April 1967 the unmanned spaceprobe Surveyor 3 landed on the Moon as a prelude to the Apollo manned Moon landings. On 12 November 1969, the Apollo 12 lander, carrying astronauts Pete Conrad and Alan Bean, touched down close by. They detached and brought back one of the cameras from the Surveyor probe. They were careful to keep it in a sterile condition but back on Earth between 50 and 100 bacteria of the *streptococcus mitis* variety were found alive inside. The survival of these bacteria was no mean feat. They had survived in the high-radiation environment of space for about three years, in temperatures that were around $-250°C$ and with no water or food source. The discovery was so amazing that in the 1980s *bacillus subtilis* bacteria were specifically put in NASA's Earth-orbiting Long Duration Exposure Facility (LDEF) and collected six years later to see how they coped. Enduring the cold and the lack or oxygen was not much of a problem for them but the exposure to ultraviolet radiation and cosmic rays (tiny bits of atoms that come hurtling through the Solar System at velocities close to the speed of light) proved more damaging.

Nevertheless, some believe that if bacteria could be incorporated into meteorites then they may be able to survive for very long periods of time in space. After all, they argue, bacteria from the gut of a bee that had been trapped in resin for 25 million years were successfully revived in a laboratory.

Perhaps this mechanism could work in some limited fashion for cross-contamination between planets in the same Solar System but I have to admit that it seems pretty difficult to conceive of a mechanism for enabling meteorites to travel interstellar distances.

So, with the greatest respect, I cannot possibly advocate support for the panspermia hypothesis, as I do not think the evidence supports such a radical claim. Anyway, it does not solve the origin of life problem; it simply moves it to a different location. Life had to start somewhere and I believe it started right here on Earth. But where on Earth?

THE SCENE OF THE CRIME

By considering the fossil evidence for life and the formation of the Solar System, it seems to me that science has established the time-frame in which life formed. It certainly happened sometime between 3.5 and 4 billion years ago and most probably happened between 3.85 and 4 billion years ago. Having decided when, we have to determine *where* the formation is likely to have taken place.

For many years it was considered that life began in pools of water containing a dilute chemical 'soup', dubbed 'primordial soup'. It was assumed that the primitive atmosphere of the Earth consisted largely of methane and ammonia. In 1953, biologists Stanley Miller and Harold Urey performed a now classic experiment. They filled a vacuum flask with an atmosphere of methane, ammonia and hydrogen. To this they attached a flask of water which they heated to produce water vapour. A third chamber was fitted with electrodes so that the water vapour and other gases could be subjected to electrical sparks to simulate lightning. A fourth chamber allowed the water vapour to condense and rain-out into an 'ocean', taking with it any chemicals that the lightning had caused to form.

After running this experiment, they discovered that three types of amino acid – the building blocks of proteins – had been produced. It appeared as if a giant leap forward had been made and the first step had been taken towards the chemical synthesis of life itself. It now seems unlikely that this was the case. It is currently thought that the primitive atmosphere of the Earth is more likely to have been predominately composed of carbon dioxide. When Miller–Urey experiments are run with CO_2, the results are far less encouraging. Also, the Murchison meteorite, a left-over remnant of planet formation, fell to Earth, in Australia, in the late 1960s. It has been found to contain many types of amino acid, not just those needed for Earthly proteins. It therefore appears that interstellar processes can synthesize amino acids and then seed the planets with them, along with the volatile chemicals needed for the atmosphere and oceans.

So the influx of comets would appear to bring a sufficiently complicated set of chemicals. The oceans can provide water to dissolve the chemicals and jostle them together so that chemical reactions can take place. Next, all we need is an energy source to set up our dissipative structures and, hopefully, life will form. The most likely energy sources lie at the bottom of the ocean.

On the beds of the oceans, mysterious objects known as 'black smokers' are found. They constantly jet streams of superheated water into the cold depths of the ocean. Typically, the water issuing from these 'chimneys' on the ocean floor is at a temperature of about $450°$ C and is laced with all manner of dissolved chemicals. The water does not boil because of the weight of the overlying water pushing down on it but, as it mixes with the frigid $10°$ C water in the surrounding ocean, the chemicals precipitate out and form the black clouds that give these objects their colloquial name. More formally, black smokers are called hydrothermal vents and are a form of volcanic activity.

A black smoker is formed when cold ocean water percolates down into the crustal rock. As it nears the molten magma below the ocean floor, it heats up and is jetted back up and out into the ocean. During its subterranean adventure, the water dissolves chemicals. At the boundary between the superheated jet and the cold ocean, an abundance of microbial life is found. These little critters love the temperature of the water and, surprisingly, could not survive in more clement surroundings. The most remarkable thing about these hyperthermophile microbes is that genetic research suggests that they are the most primitive life-forms on Earth.

Literary historians can trace the early history of Geoffrey Chaucer's fourteenth century work *The Canterbury Tales*, by comparing the differences between copies of the text. In those days, scribes would copy the manuscript by hand. Whilst they would copy the manuscript as faithfully as possible, it is obvious from a comparison of any two copies that every now and again they simply could not help themselves and rewrote small pieces of Chaucer's deathless prose. When the transcribed version was subsequently copied, not only would the changes be copied but new changes would also creep in. So with time, the text diverged from the original. There are 58 known surviving mediaeval manuscript copies of *The Canterbury Tales*, all of which have obviously been changed from the original – some by more than others. By comparing the textual differences, scholars of literature have been able to deduce a family tree that allows them to determine which manuscripts are the closest to Chaucer's original.

Biologists can do something very similar with life-forms. They choose a specific strand of messenger RNA that is common to all life: from the humblest microbes to humans. Genetic mutations have crept in over time and this allows biologists to compare the mRNA base sequences and deduce the tree of life. The hyperthermophiles sit closer to the bottom of the tree than anything else. However, they are not the common ancestor of all life. The same research makes it clear that the eukaryotes diverged before the present-day hyperthermophiles developed and whatever came before them has long since died out and left little or no trace. Nevertheless, it seems very probable that the common ancestor of life on Earth was hyperthermophilic too and originated in the vicinity of the black smokers. Imagine that! A bug that once lived at the bottom of the ocean in scalding water was your great-great-great-great-great ... grandmother!

The more I think about it, the more it seems to me that the hydrothermal vents are the location of the origin of life. They seem ideally located for such an event. Under the ocean they would have been protected from the occasional stray comet that hit the Earth during the latter period of the late bombardment. The flux of hot water and chemicals provide a perfect environment for open thermodynamic systems. The only trick remaining is the development of some kind of chemical dissipative structure.

THE CHICKEN AND THE EGG

The origin of life has long been thought of as the ultimate 'chicken and egg' situation. This particular old puzzle is: 'If you need a chicken to lay an egg but you need an egg to create the chicken to lay the egg, which comes first?' The question is insoluble.

The problem with living cells is that the nucleic acids and proteins are so wrapped up in each other's continued survival, it is similarly almost impossible to determine which came first. For example, a DNA molecule can only reproduce when an enzyme pulls it apart and uses it as a template to assemble a DNA molecule. The enzyme necessary to do this is a protein which is built because a gene, contained in the DNA strand, says it should be. It is the 'chicken and egg' situation all over again.

Stuart Kauffman, of the Santa Fe Institute, believes he has the answer. He has built on the work of German biophysical chemist, Manfred Eigen. Using computers, Kauffman has modelled sets of chemical reactions. As more and more reactions take place, building up more and more complicated molecules, some of these molecules will be able to act as catalysts, speeding up certain other reactions. Now, what if a catalyst speeds up a reaction that eventually leads to its own production? That sequence of chemicals becomes an autocatalytic set. In effect, the catalyst molecule helps its own creation. Effectively the molecule becomes capable of self-reproduction – one of our properties of life. Networks of ever more complicated molecular interactions will form, with the most efficient catalytic networks producing the most molecules. These will tend to be the networks with the smallest number of steps between the catalysed reaction and the production of the catalyst.

Next, think about DNA. Because of the specific pairing on nucleotides, to make an exact copy of itself, all that a strand of double-helix DNA needs to do is be unwrapped and let other nucleotides bond. There is nothing more efficient than this one-step process. As soon as a molecule develops some kind of specific pairing, it will rapidly dominant the other reactions. Because it came from an ever increasingly complicated reaction network, there are likely to be all sorts of other reactions in which it inevitably takes part. So the concept of the genetic chicken and egg is solved. Both came first – proteins *and* nucleotides. No single specific molecule is important, so the chance of a DNA strand forming at random is meaningless in the origin of life debate. Instead, it is the reaction sequence that is the key – and these develop naturally, as dissipative structures, whenever there is an influx of energy and food chemicals to drive the reactions. It is these naturally occurring reaction systems that are governed by a chemical 'survival of the fittest' approach, with the most efficient system rising to dominate all of the others.

It is specific pairing which gives a molecule the most efficient system of self-reproduction and it is also specific pairing that allows genetic information to be passed from molecule to molecule. Life, I feel, cannot help but follow.

With this viewpoint we should expect life to 'crystallise' again and again, providing that the chemical system is sufficiently complex to take us to the edge of chaos but not too complex that it plunges us over the edge. Each time the 'life' may be based on different molecules and copying processes but it will be off the starting blocks and running.

From here on in, evolution is your guide. Now it is time to think a little more about what constitutes a habitable planet because, if the origin of life is as inevitable as I have suggested in these last few pages, then any planet that is habitable will also be inhabited.

Chapter Six

Sites for life

B elieving that life is not an accident implies that I also believe that any
planet *capable* of being inhabited *will* be inhabited. As Jules Verne
wrote in his book *From the Earth to the Moon*, published in 1877: 'I
should venture to assert, that if these worlds are habitable, they either are,
have been, or will be inhabited.' So for the rest of the book I'm with Jules on
this one and the term 'habitable planet' will be used synonymously with
'inhabited planet'.

As I stated in the last chapter, the conditions on Earth that heralded the
origin of life were not the same as those of today. Habitability is certainly in the
eye of the beholder. A methane-breathing microbe that lives a few kilometres
under the surface of the Earth in solid rock is not going to be impressed if I
simply state that an oxygen-rich atmosphere equals a habitable planet. During
this book so far, I have mentioned, in passing, some bizarre microbes that can
withstand incredibly extreme conditions. I can no longer delay their
appearance centre stage. Ladies and gentlemen, please welcome the current
stars of the microbial world: the amazing extremophiles!

EXTREME ENVIRONMENTS AND LIFE

Despite sounding like a 1970s pop group, the extremophiles are actually micro-
organisms that inhabit places that you and I would find totally unbearable –
much worse than a rainy weekend in Clacton-on-Sea. Yes, that bad. Some
extremophiles like it hot, some like it cold, some like it deep and some like a
cocktail of noxious gases to breathe. During the last few years, all of them have

rocked our former concept of habitability and forced a revolution in thinking. They have focused our attention on just how diverse habitable niches can be on the Earth and made us look at the other planets in the Solar System in a new light.

Surprisingly, the basic knowledge of these organisms' existence is nothing new. For 80 years, scientists have known of peculiar microbes that live quite happily in saturated salt solutions. Then, in 1952, the Danish Galathea expedition discovered other microbes that like the high pressure and extremely cold environments found at the bottom of the deepest ocean trenches in the world. The 1970s brought with them the realisation that, at the other end of the scale, temperatures close to boiling, as well as highly acid or alkaline water could also host micro-organisms. In the decades since, the upper end of the habitable temperature scale has risen to 114° C (under the high-pressure conditions that prevent the water from boiling).

The reason for their sudden increase in status is because scientists now recognise that the hyperthermophiles are vital to our understanding of the origin of life – as I mentioned in the previous chapter. It has been estimated that probably less then 0.1% of micro-organisms have been discovered. I cannot help but wonder what amazing creatures remain to be found here on Earth. Could the boundaries for life extend even further?

Researchers studying the hydrothermal vents in the Pacific Ocean, close to Tahiti, may have pushed the boundaries further than anyone could possibly imagine. The pilot of a University of Hawaii submersible saw a bizarre jelly-like substance issue from the Loihi vent. The appearance of the strange material was reminiscent of that of some deep-sea fish. Could it possibly be alive when temperatures in the vent can be higher than 400° C? Researchers are currently busy trying to find out.

With the extremophiles in mind, we need to keep our definition of habitability as loose as possible. This is great news because, if the evidence were pointing to a strictly defined set of prerequisites, the chances of life appearing elsewhere would diminish. The more examples of extremophile life we find on Earth, the greater the chances of there being life in alien environments on other worlds.

What might be the requirements for life? Even extremophiles need water, so I will state that the first requirement is that the world in question possesses liquid water. This leads us to the concept of the habitable zone. This is an easy concept to grasp but is somewhat harder to pin down mathematically. Simply think of the habitable zone as being a region around a star in which a planet will be warmed sufficiently by the star's radiation that liquid water will be present on its surface.

REQUIREMENT 1: WATER, WATER EVERYWHERE

I emphasised the importance of liquid water during the discussion of the origin of life but the story does not stop there. If our world were to dry up tomorrow, it would not only be mineral water companies that noticed. All life on Earth is based upon the availability of liquid water. This is because the chemical reactions that sustain life are performed in a liquid-water solution. An alien character in *Star Trek: The Next Generation* once described humans as something like 'ugly bags of mostly water'. Whilst I am understandably reluctant to discuss the attractiveness of the show's cast members, I am prepared to admit that the alien was right about our being mostly water.

In fact, about 60% of a human body is water. It provides a liquid base to dissolve nutrients and to allow reactions to take place. In a gas, molecules are too far apart to easily react and in a solid, molecules are generally held too rigidly to interact. So life is always going to be based on a liquid and on Earth, life simply would not be possible without water in liquid form.

When it comes to defining the habitable zone in terms of distances from the central star, things start to become complicated. Imagine the Earth as a ball of rock in space, with no atmosphere. As it orbits the Sun it will be bathed in sunlight and warmed. At a distance of 1 AU, the Earth would receive enough energy to heat its surface to a temperature of just $-18°$ C. This is not good news if you want to find liquid water.

Luckily, the Earth is not just a rock in space; it is a rock in space with an atmosphere. We have all heard about the greenhouse effect and I am willing to bet that most people believe that it is a bad thing and to be avoided at all cost. The amazing truth is that without the greenhouse effect, life on Earth would be impossible because it would be too cold. The principal greenhouse gases are carbon dioxide and water vapour. They work by allowing sunlight into the planet's atmosphere, where it can strike the ground and warm it. At night-time, the warm ground re-radiates this energy but at longer, infrared wavelengths. The blanket of greenhouse gases traps this heat and does not let it escape back into space, thus keeping the planet's temperature higher than it would otherwise be.

Venus is often described as a twin of Earth and in terms of size and mass this is almost true. Venus is only slightly smaller than the Earth. At a distance of 0.723 AU, it receives 1.9 times more sunlight for every square metre than does Earth. This has triggered a runaway greenhouse effect that now bakes the surface of the planet to a hellish $475°$ C.

Mars, on the other hand, could use a thicker atmosphere to keep it warmer. At a distance of 1.52 AU, Mars receives only about 0.43 times the sunlight per square metre than Earth receives. It has a tenuous atmosphere of carbon dioxide that is incapable of raising the temperature very much. Variability in the atmosphere means that temperatures on Mars are usually well below the freezing point of water but, occasionally, it can rise to $0°$ C – or just about. This provides us with the hope that life might just be hanging on in a few sheltered martian locations.

So the habitability of a planet depends not only on the distance of the planet from a star but also on the thickness of the planet's atmosphere. In the last chapter I explained that an atmosphere is also necessary to provide the downward pressure to create liquids and stop a melting solid turning immediately into a gas. Yet another consideration is that surface-dwelling life will need something to breathe. So whichever way we look at it, an atmosphere is vitally important for a habitable planet.

All other things being equal, it is the mass of the planet that determines its gravitational field. Since gravity is responsible for holding the atmospheric gases to a world, this must set a lower limit on the mass of a habitable planet. Compare the Earth and the Moon; one is habitable and the other is not. The Moon is simply too small to possess enough gravity to retain an atmosphere.

The smallest body in the Solar System to possess a substantial atmosphere is Saturn's largest moon, Titan. It is 5,150 km in diameter, compared with the 3,476 km of our own Moon. Titan is surrounded by a dense atmosphere of methane and nitrogen but is probably near to the lower limit for an atmosphere-bearing world. This is because other moons of similar size – Ganymede and Callisto, for example – do not have appreciable atmospheres. In general, the larger the world, the more easily it retains an atmosphere.

If the planet is too big, however, it retains too dense an atmosphere. The ideal mass range for a habitable planet is somewhere between Titan's 0.02 Earth-masses and 2 Earth-masses. Greater than this, the atmospheric pressure would become too dense to allow complex Earth-like organic chemistry.

Habitability is not just dependent on the planet. The star too has its role to play because there are different types of star in the Universe. I mentioned some of them in Chapter 3, when I discussed high- and low-mass stars. Astronomers classify the vast majority of stars into one of seven different categories that are identified with one of the following letters: O, B, A, F, G, K, M. Male astronomers coined the phrase 'Oh, Be A Fine Girl, Kiss Me,' to remember this sequence; female astronomers can replace 'Girl' with 'Guy'. The ladies can also be smug about the fact that it was two women who improved the classification scheme originally devised by men. It had run A, B, C ... and so on, to P, with the classification Q for anything they could not make up their minds about. Antonia Maury and Annie Jump Cannon pointed out the errors of the male way (well, those associated with the classification scheme anyway) by showing that most of the gentlemen's categories were unnecessary and, of those that remained (A, B, F, G, K, M, O), the order was really O, B, A, F, G, K, M when the surface temperature of the stars was taken into account. O are the hottest at over 20,000 K and M are the coolest at about 3,000 K.

These differences in stellar temperature will obviously influence planetary temperatures. If the Sun, at 6,000 K, were to be replaced by an O star at 20,000 K, the surface of our world would be fried to a crisp. At the other extreme, replacement by an M star would render our planet a frozen ball of ice.

So, the presence of liquid water on the surface of a planet is dependent upon the following three factors:

The type of star around which the planet is in orbit.

The distance of the planet from the star.

The planet's atmospheric density and composition.

Then, just when you thought you had it all figured out, along comes the realisation that the energy a star releases is not constant over billions of years. During its lifetime a star becomes progressively hotter as helium builds up in its core after the fusion of hydrogen. Hold that thought for a moment, as I will return to it shortly, during the section on how stable an environment has to be to allow life to flourish. For now, let us ask the perfectly valid question: 'Could another liquid replace water?

ALIEN CHEMISTRY

To be worthy of consideration, the substitute liquid would have to be naturally occurring and approximately as abundant as water is on the Earth. It may, however, be in a liquid state across a different temperature range. Two possibilities immediately spring to mind. They are other astronomical ices: ammonia and methane.

The temperatures required for these to be liquids, along with some other substances that should not be discounted, are shown in the table on p. 81. As can be seen, the temperatures required are all pretty chilly; so a living system based on one of these liquids would need to be on a planet much further away than the Earth is from the Sun.

Fascinatingly, Saturn's moon Titan is thought to have a surface temperature of 94 K, so methane or ethane could exist in liquid form on its surface. The Huygens probe, which will plunge into Titan during 2004, has been specially designed so that it will float if it encounters a methane or ethane ocean.

The only problem is that when the temperature drops, so does the rate at which chemical reactions take place. Typically, if the temperature drops by

Liquid range for various chemicals

Liquid	Freezing point	Boiling point	Liquid range
Water	0° C	100° C	100° C
Ammonia	−78° C	−33° C	45° C
Methane	−182° C	−164° C	18° C
Ethane	−183° C	−89° C	94° C

10° C, the reaction rate is halved. On Titan, which is about 200° C colder than Earth, the chemical reaction rate would be over a million times slower. Evidence on Earth suggests that life originated within the 500 million year window between 3.5 and 4 billion years ago. If the reaction rate is slowed by a factor of a million, life on Titan might not be expected to begin within 5×10^{14} years – a hundred thousand times longer than the current age of the Solar System.

Before we write off Titan, however, remember the enzymes: those proteins that speed up reaction rates by acting as catalysts. If it were not for these catalysts, Earthly biochemistry would slow to a crawl: a lungful of air would take hours to be metabolised and a meal would take several weeks to be digested. The energy release would therefore be slower and what we would usually do in a few hours would take us weeks to accomplish. Eureka! Perhaps I have just discovered why construction work often takes longer to complete than anyone thinks it will.

Returning to the point, it is conceivable that, as one of Stuart Kauffman's autocatalytic reaction networks takes affect, enzymes speed up the reactions at an ever accelerating rate. As much as I like this idea, I am not holding my breath for a momentous discovery of life when NASA/ESA's Huygens probe reaches Titan in 2004.

A potential problem for some of the alternatives is that the temperatures that allow them to be liquid are quite restricted to a few tens of degrees, which means that just small changes in the climate might be enough to freeze or boil the planet and bring life to an end. Although water can do little to stop itself boiling, it does have a unique property that in certain situations can protect it against freezing: when water crystallises to become ice, its density decreases. In other words, the amount of ice that can be placed in a container is smaller than the amount of water that the same container could hold. This means that when water freezes, the ice floats. Anyone who watched the film *Titanic* knows this to be true. No other common, naturally occurring liquid does this.

The reason for the importance of this is that when a lake begins to freeze, an ice skin develops over the top and insulates the water underneath, allowing it to remain liquid for longer. Thus, life in the lake can go on as normal. If the lake were to freeze completely, the water in the living cells of the lake's inhabitants would freeze. The water would expand to become ice and burst open the cell membranes, killing the cell and therefore the organism. This is one of the greatest problems facing scientists who dream of creating hibernation chambers for humans: the act of freezing a body destroys its cells. There are some extremophile micro-organisms that sidestep this problem by having a kind of natural anti-freeze but, in general, $0°$ C spells 'G-A-M-E O-V-E-R' for life.

All our alternative liquids become denser when they freeze and they therefore sink. The up-side of this is that if an organism based on another liquid were to freeze, its cell contents would not expand and rupture their cell membranes. So these alien creatures might be able to survive long spells in 'suspended animation' and be capable of carrying on where they left off once they thawed out again. This may also mean that the restricted liquid ranges of these other chemicals are not so much of a problem after all. As with all speculation, however, until life based on these other liquids is found, we simply do not know.

REQUIREMENT 2: THE RIGHT CHEMICALS

It seems obvious to say this but without organic chemicals – those containing carbon – life on Earth would be impossible. So without the right chemicals, a planet cannot be habitable. The trick is in knowing the right chemicals. I presented an inventory of the most common cosmic chemicals in the table on p. 38 and in the same chapter, I discussed the way they are formed in the hearts of stars. The question now is: 'Does life have to be based on carbon?'

We say that life is based on carbon because every molecule in our body contains one or more carbon atoms. The easiest way to imagine life based on another chemical is to think of an atom with approximately the same characteristics as carbon. Is it not then reasonable to suppose that it could perform the same function as carbon?

This train of thought has led to the speculation that silicon might be a life-giving atom because both silicon and carbon can form molecules that are essentially long chains of atoms. With carbon, this allows the building of proteins and DNA. Silicon chains are weaker than carbon chains and can be more easily broken apart. There is also the problem of incorporating silicon into life-forms under Earthly conditions. Carbon dioxide dissolves in water and the carbon can be incorporated into the organic substance of an organism that breathes carbon dioxide. Plants, too, breathe carbon dioxide, extract the carbon and combine it with water, giving off 'waste' oxygen in the process. Therefore, the entry of carbon into the food chain relies on its presence in the form of gas. There are no molecules containing silicon that can exist as a gas under normal Earth-surface conditions. Although it seems that carbon is the element best suited to life on an Earthly location, silicon cannot be completely written off. Alien conditions – such as very different temperature ranges or different pressures – might mean that silicon behaves in a way more conducive to living molecules than would carbon under those particular circumstances. What would these conditions have to be? The answer might be below our very feet. Iconic scientist, Thomas Gold, has suggested that the deep interior of the Earth – where conditions are so extreme that no one has been able to simulate them in the laboratory – may promote the existence of silicon-based life.

Could it be that the Earth possesses two biospheres? One on the surface, with which we believe we are familiar and a deep hot biosphere that remains to be discovered? It is a tantalising prospect if, at present, an eccentric one.

REQUIREMENT 3: AN ENERGY SOURCE

Despite the fact that earlier in the book we relegated the energy radiated by a star to be the waste product of its formation, in the case of the Sun, it is essential for the continued survival of a large proportion of life on Earth. There was once a time when the Sun was said to be the energy source of all life on Earth. This is no longer the case.

The communities of microbes and larger sea creatures that live near hydrothermal vents on the ocean floor owe nothing to the Sun. If it stopped shining tomorrow it would not bother them because the Sun's rays penetrate only the first 200 metres into the ocean. Within this zone, photosynthesis by marine organisms can take place. On the ocean floor, however, that is one luxury not open for exploitation. Instead, they must rely on the geothermal energy being pumped into the ocean by the black smokers.

In Jupiter's system, another form of energy is available in sufficient quantity to have a profound effect on some of the moons around the giant planet: tidal energy. Most of us are familiar with tides but perhaps associate them only with the ebb and flow of an ocean. To an astronomer, a tide is not necessarily manifested by water; it is caused by differences in gravity whenever one celestial object is in orbit around another. In this configuration, the gravitational force keeping one object orbiting the other is greater on its near side than on its far side. This tends to pull the celestial objects into the shape of a three-dimensional ellipse rather than into a sphere (see the illustration on p. 84).

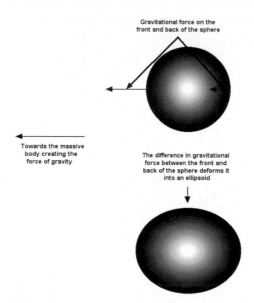

Tidal forces. The difference in gravitational pull across an object causes it to be distorted from a sphere into an spheroid: a three-dimensional ellipse.

In the case of the Earth and the Moon, both objects raise tides on one another. On Earth, the water in the oceans responds readily to the Moon's gravity and so changes its position relative to the land by about 1–2 metres in altitude. These changes manifest themselves as the familiar tides.

Around the giant planet Jupiter there is a collection of moons, four of which were first seen by Galileo in 1610. These Galilean satellites are called Io, Europa, Ganymede and Callisto, in increasing distance from the planet. Io and Europa, the two closest moons, suffer spectacularly from tides. As Io sweeps around its orbit, the tides squeeze it so much that friction melts its entire interior and sulphurous lava is constantly spewed through volcanoes onto its surface. In fact, Io is the most volcanically active body in the whole Solar System. It is never in a state of anything less than catastrophic eruption (see the illustration on p. 85).

Europa, the next moon, is far more important in terms of habitability. The same tides that torture Io also have Europa in their grip but because Europa is further from Jupiter (670,900 km, compared with Io's 421,600 km) the effects are not as severe. Even so, the surface of Europa is raised by some 30 m because of the tidal force. Most of Europa is thought to be a mass of solid rock, surrounding a metal-rich core. The Voyager spaceprobe, which flew by Jupiter and its moons in 1979, provided tantalising evidence that, below the icy crust, Europa possesses a global ocean of water, 100 miles deep – about 20 times deeper than Earth's oceans – kept liquid by tidal friction caused by Jupiter. The

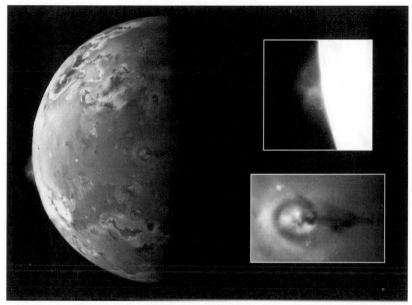

An eruption on Io. This moon of Jupiter is the most volcanically active body in our entire Solar System. The tidal force of Jupiter and the fact that Io is in a slightly elliptical orbit, causes the moon to be squeezed. This creates friction that entirely melts its interior and causes the volcanic activity. (Photograph reproduced courtesy of NASA, JPL, Galileo Project Team.)

Galileo spaceprobe of the late 1990s has studied the jovian system and has paid special attention to Europa. It has returned images and data that strongly suggest the presence of liquid water below its surface. Some of the pictures show amazing ice floes and features that resemble volcanic structures on the floors of Earth's oceans. In those locations, molten rock rises from the interior of the Earth and, as it makes contact with the water, solidifies to form new oceanic crust. On Europa the ice plates move apart, allowing liquid water to well up and immediately freeze, creating new areas of ice crust (see the illustration on p. 86). This behaviour would be next to impossible if it were not for the fact that ice floats on water.

There is a faint possibility that regions of the subsurface of both Callisto and Ganymede may have pockets of water. However, Europa is the undoubted star of this show and plans are already afoot to further investigate this particular world – as I shall explain in the next chapter.

If life *did* begin at the bottom of the ocean, driven by hydrothermal energy, then volcanic activity on a planet may prove to be the essential energy source for the establishment of life and perhaps even for its continued survival. It is a virtual certainty that as the planets formed they were hot. This is because they

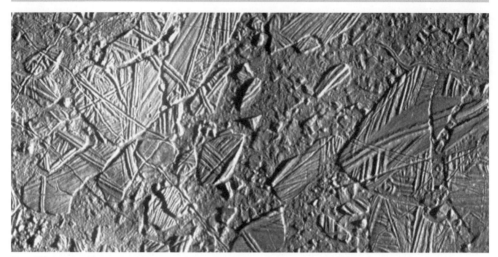

Ice Floes on Europa. Unprecedented images from the Galileo spaceprobe have shown fractures and floes in the icy crust of this moon. There are strong indications that a global ocean exists under the surface. (Photograph reproduced courtesy of NASA, JPL, Galileo Project Team.)

accumulated by the collision of smaller planetesimals. The collisions would have transformed kinetic energy into heat energy. In the semi-molten state of the planet, the dense metallic substances would sink to the centre and the lighter rocky material would float to the top. The friction caused by this process would have contributed to more heating. There would also be some radioactive elements trapped inside the forming planet. As they decay – a process that is still continuing today – they will continue to heat the interior of the planet.

The Earth has yet to radiate away all of its own heat. In particular, seismic readings indicate that the core of our planet is still partly molten. As this material freezes it gives off heat which helps keep the rocks below the crust of the Earth molten. As a result, volcanic activity is still very much a feature of our world. Both Venus and Mars display volcanoes, showing that volcanic activity was important in the past. On Mars the volcanoes are no longer active but on Venus, on the other hand, there may still be residual volcanic activity.

REQUIREMENT 4: A STABLE ENVIRONMENT

Our final requirement is pretty self-evident: the planet needs to possess a stable environment. One of the best ways to create a stable environment is to place a planet in a circular orbit around a star that is not prone to occasional flare-ups. This ensures that the planet retains approximately the same temperature throughout the year. The Earth follows a nearly circular orbit. The seasonal changes of temperature in some countries on Earth are caused by the tilt of the Earth's axis. An elliptical orbit creates large variations in temperature. Mars' two-year orbit is more elliptical. It creates such temperature differences that, during the martian winter, the atmosphere freezes at the polar caps. The first spring rays of Sun cause the ice to melt and, because there is no atmosphere to keep it liquid, it sublimes straight into a gas and reinstates the red planet's meagre atmosphere. In the process it raises enormous dust-clouds that can blanket the entire planet for weeks at a time, preventing most of the sunlight from reaching the surface.

All this makes Mars seem quite inhospitable but many scientists have high hopes of finding life there. The redeeming feature of the planet is that at least its atmospheric instability is cyclic. Every winter the atmosphere freezes out – and perhaps evolution can adjust to handle such cycles. The same applies for changes that take a long time to occur. They provide evolution with the time it needs to breed adapted life-forms.

Our geological record offers clear evidence that the Earth has changed throughout its 4.6 billion-year history and that life has managed to adapt. In fact, it may be the case that a totally stable environment is not as desirable as at first appears. This is because, as soon as creatures become optimised for their surroundings, they stop evolving because any mutation confers a disadvantage trait. A little instability to shake things up can keep evolution rolling.

In a later chapter I shall consider the evolution of life from simple creatures to intelligent beings. I will unfold a remarkable story that shows how evolution progresses in fits and starts and may actually need Earth-shaking disasters, catastrophic extinctions and short, sharp climatic changes in order to breed intelligence.

At the moment, however, it is time to consider the concepts of habitability that I have introduced in this chapter and look with new eyes at the other worlds of our Solar System. Do any of them offer habitable niches? What about the newly discovered planets around other stars; can they, too, be habitable? How would astronomers and biologists detect life on other worlds, even if it were there? The answers are ingenious, often surprising and – best of all – are presented in the next chapter.

Chapter Seven

Looking for life in the

Solar System

D'Arcy Wentworth Thompson was a Scottish zoologist who was born in 1860 and died in 1948. During his life he became Professor of Biology at Dundee University and Senior Professor of Natural History at St Andrews University. In 1937 he was knighted for his outstanding contribution to science. His masterpiece was a book, published in 1917, entitled *On Growth and Form*, in which he applied mathematics and physics to the understanding of the patterns and shapes seen in nature. He believed that the natural world was rooted in the laws of physics and that by explaining some of the elaborate structures found, humankind's understanding of life and biology would progress.

Despite his public accolades, D'Arcy Thompson's ideas on life were regarded by his peers as those of a maverick and his studies were not taken up. As the third millennium dawns, his time has come and his outstanding contribution to our understanding and recognition of life is finally being acknowledged. In fact, the techniques he advocated will be used by the most powerful computers in the world to search for life on other planets.

The endeavour is known as the Book of Life project and is being run by scientists at the Marshall Space Flight Center, using a computer technique known as a 'neural network'. This makes the computer work in a very different way from the computers that sit on most people's desks and run Windows. Normally, a PC operates as a form of sophisticated calculating machine which uses memory and processing power in regimented, analytical ways to produce incredibly precise answers to mathematical or logical problems. A neural network is far more flexible and can deal with uncertainties and with grey areas by mimicking some of the attributes of the human brain. Furthermore, it cannot simply be programmed: rather, it has to be taught. With each lesson, the neural network alters its internal settings, until it becomes an expert at the task in hand. The old adage, 'practice makes perfect', now applies to computers.

The big advantage of neural networks over ordinary computers is that they can develop the skill of judgement and make informed guesses about what the neural network is being asked to interpret. In Chapter 4, during our agonising over a definition for life, I said that life is easy to recognise but much harder to define. If we could define it by a set of simple rules, we could devise a test and programme a traditional computer. Since we cannot, we must use a neural network and train it to recognise life.

The neural network of the Book of Life project is based on a super-computer called Leibnitz, after the German mathematician whose lifelong goal was to organise the entire sum of human knowledge. Using an electronic camera as the computer's eye, the scientists working with Leibnitz have trained it to recognise the basic microbial shapes: spheres, rods, filaments, cocci (cluster of microbes that resemble bunches of grapes) and spirochete (spirals). It has so far classified more than 10,000 Earth microbes and is constantly improving its ability to recognise them.

It is hoped that the Book of Life neural network will be so sharp at spotting life that samples of material from Mars, or wherever, can be passed before its electronic eye and that it will quickly and efficiently tell scientists whether anything in the sample looks as though it might be – or once was – alive.

Because all of the microbes used to teach Leibnitz have grown on Earth, they are subject to the physical force of Earth's gravity. Eventually, however, Leibnitz will be so clever that if it is told that the samples came from Mars, it will know that Mars' gravity is just one third of the Earth's gravity and therefore will adjust its expectations of microbial shape.

Leibnitz is the high-tech tip of an enormous exobiology (the study of life beyond the Earth) effort. Research teams from all over the world are now dedicated to finding life in the Universe, or, at the very least, evidence of past life. Amazingly, it is the worlds of our Solar System that are being most closely scrutinised.

MARTIANS GO HOME!

At various times throughout history, the scientific community and the general public have believed in the wide-scale presence of intelligent life on Mars – some people still *do* believe it. I, however, do *not* and neither does the overwhelming majority of scientists and the general public.

Prior to the 'discovery' of fossils in the martian meteorite ALH84001 – which soon enough will rear its ugly head in this book – the most famous martian-life episode began in 1877, when Giovanni Schiaparelli observed the red planet during its close approach to Earth. He drew maps of what he saw, which consisted of a number of straight lines criss-crossing the planet. He called them *canali* – Italian for channels – but a mistranslation into English resulted in their forever being called 'canals'. At the time, the Earthly canal system was the pinnacle of engineering prowess. Suddenly, it was felt that clear evidence of a Martian civilisation had been found. Astronomers had visually identified the polar caps of Mars and some folks wondered whether Mars was a dying world – perhaps the canals were the last, desperate attempt of the planet's inhabitants to irrigate the parched equatorial regions ...

Before you could shout 'my favourite Martian!', master science-fiction writer H.G. Wells had written the epic *The War of the Worlds*, in which the Martians give up the fight for their own world and instead come to conquer Earth. Luckily, the Martins succumb to all sorts of nasty Earthly bacteria and it serves them right. Alien bullies in three-legged fighting machines deserve no better, if you ask me.

As more people observed the planet Mars, some saw the canals and some did not. The maps of the canals were a mystery because no two ever looked the same. What were those pesky Martians up to? Eventually, the reason for the discrepancy was realised: the canals were optical illusions. The finest telescopes of the day could *almost* resolve the surface features – but not quite. As the Earthly human brain interpreted this marginal data, it caused the lines to be 'seen'. Today, small telescopes – the equivalent of the finest Schiaparelli possessed – can be used to see 'canals' on Mars!

Nowadays, when you hear talk about the search for life on Mars, what is being referred to is the hunt for microbial life at best or, at the very least, fossils of long-dead creatures. The reason for the increased belief that Mars was an inhabited planet at some stage in its history, is that the continuing exploration of the planet, by spaceprobes and robotic landers, is providing compelling evidence that Mars was once a world on which liquid water flowed.

Planetary geologists can see many examples of rock features on the red planet that suggest the widespread distribution of water in the past. The northern hemisphere of the planet is much lower than the southern highlands and some suggest that most of the lower-lying half of the planet was once covered by an ocean. The idea of the widespread distribution of liquid water on the martian surface is known as the 'warm, wet' hypothesis and relies on the fact the Mars was volcanically active during its youth. A rival hypothesis is the 'cold, wet' Mars, suggesting that enormous ice-sheets once covered the planet and that because of the insulating effects of ice, large quantities of water could still flow beneath the ice.

For the primitive martian climate to be warm and wet, the planet needed a substantial carbon dioxide atmosphere to produce a large greenhouse effect. Remember that Mars is 1.5 times further from the Sun than is the Earth and so receives less solar radiation to heat it. As if this were not bad enough, the Sun would not have been as bright billions of years ago as it is today and this would also limit the effective heating. Even a cold, wet Mars would need some level of greenhouse effect.

The evidence for past volcanism on Mars is good and tends to favour the warm, wet theory. It is a virtual certainty that on Mars today, volcanic activity is dead, or, at the very least, highly restricted. The enormous but extinct martian volcano, Olympus Mons, is the largest volcano in the Solar System. In 1999, an unprecedented discovery by the NASA spaceprobe, Mars Global Surveyor, provided surprising evidence that the level of past volcanic activity on Mars was substantially higher than had been previously thought. In fact, it now seems likely that a process known as 'plate tectonics' was active on Mars in the past, making the planet substantially more Earth-like during its early stages.

Plate tectonics is the mechanism by which the crust of our planet 'floats' on molten rocks in the mantle. Convection in the mantle moves the crustal plates, which causes volcanoes, earthquakes and creates mountain ranges. The volcanoes are particularly important because they release gas that has become trapped in the planet's rocks back into the atmosphere. The whole system of plate tectonics on Earth is caused by our planet's release of internal energy. Stop that energy and the plate tectonics will stop. The volcanoes will die and the atmosphere will lose one of its maintenance facilities. On Earth, life is so widespread that it provides a means of atmospheric replenishment. In fact, the algae that live in the upper layers of the oceans are described by independent scientist, James Lovelock, as the 'true lungs' of the Earth.

Plate tectonics causes the sea-floor spreading that I briefly described in the last chapter when talking about fractures in Europa's ice crust. As the magma wells up and solidifies to make new oceanic crust, the magnetic field of the Earth imparts a magnetic field onto the rocks. Study of the magnetic field of the rocks on the ocean floor reveals that sometimes the Earth's magnetic field has pointed one way, whilst at other times it has flipped over to point the other way. There have been nine reversals in the past four million years. At this very moment, the south magnetic pole of the Earth is in the northern hemisphere; therefore, all compass manufacturers lie! Their compasses point to magnetic south, which happens to lie close to geographic north.

Mars Global Surveyor has found similar magnetic strips in rocks on the planet's oldest terrain. This is a big clue that plate tectonics was once an active process on Mars and that the planet contained a strong magnetic field rather than the weak residual field of today. Both the magnetic field and the plate tectonics make the planet more Earth-like and raise the possibility that volcanically driven hydrothermal vents in a martian ocean may have helped life to begin there.

THE AGES OF MARS

The past 4.6 billion years of martian history can be split into three geological ages; the Noachian, the Hesperian and the Amazonian. They are based on the supposed ages of rocks found in three regions of the planet: Noachis, Hesperia Planitia and Amazonia Planitia. Planetary geologists have further subdivided these into early, middle and late periods but, until rocks from these regions can be brought to Earth and dated accurately, there is still uncertainty about the actual duration of these periods (see the illustration on p. 93).

During the oldest of these three periods – the Noachian (4.6 to 3.5 billion years ago) – the conditions on Mars are thought to have been conducive to the formation of life. There was widespread volcanic activity, a strong magnetic field and running water. During the equivalent period in the Earth's history – close to the boundary of the Archean and Pre-Cambrian period – life started on Earth.

During the next martian period – the Hesperian (3.5 to 1.8 billion years ago) – the planet began to change. I suspect that one of the major reasons for this change was that the internal heat source was closing down. Being smaller than the Earth, Mars would not be expected either to contain as much radioactive material or to take as long to release all the heat trapped during its formation. As the martian iron core solidified, so the magnetic field collapsed to its present weak state. This allowed high-energy particles from the Sun to enter the atmosphere and permitted them to begin eroding it away. The process is known as 'sputtering' and takes place when fast-moving fragments of atoms collide with atmospheric gases. The volcanic activity ceased to be widespread and so did not recycle much gas back into the atmosphere. As the atmosphere was inevitably lost, so the planet gradually became colder.

Another effect was taking place that led to a substantial reduction in the amount of water on the planet. It is related to Earth's ozone layer, which we

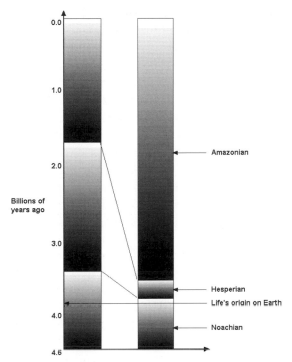

Geological ages on Mars. Two competing theories for the geological history of Mars are compared. Importantly, for both theories, water was available on the planet's surface at the same time that life began on Earth. Could life have also started on Mars?

hear so much about. It may come as a surprise but the oxygen atmosphere of the Earth was the first example of irrevocable pollution by this planet's inhabitants. Oxygen was the waste product of many ancient microbes. In effect, they poisoned their own air. Through chance mutations, however, a strain of microbe that could breathe oxygen evolved and led to the animal kingdom of today. The build-up of oxygen in the atmosphere promoted the development of the ozone layer. Ozone is a molecule made of three bonded oxygen atoms. It is a very good barrier, preventing the Sun's ultraviolet rays from penetrating the lower layers of the atmosphere where they can damage living cells and break up molecules such as water. (Think of ultraviolet radiation as being the guys you do not mess with because you are bound to come off worse.)

The fact that Mars did not develop an ozone layer may be seen as evidence that wide-scale life never flourished on the planet. Or, perhaps, the removal of the magnetic field caused a fledgling martian ozone layer to be eroded away by sputtering. In any case, when the water molecules were destroyed, the two hydrogen atoms were light enough to escape into space. The oxygen atom sank to the surface of the planet and combined with the iron in the rocks to form rust. Most people wonder why the red planet is red. The amazing truth is that it is rusty world.

Water that was not destroyed in the atmosphere must have been trapped underneath the surface of the planet as permafrost, where it still resides today. Just how much water is locked up below the surface of the planet is hotly debated. Some even believe that it would be sufficient to reinstate the ocean in the northern hemisphere.

By the time we arrive at the Amazonian period (beginning at 1.8 billion years ago and continuing to the present), the vast bulk of the martian atmosphere has been lost, liquid water has all but disappeared from the surface of the planet and volcanism has died. The dynamic early Mars has given way to the Mars of today: cold, barren and, on the surface at least, very dry.

If life started on Mars during the Noachian period, the question is, could it have evolved to compensate for the changes of the Hesperian period and therefore eked out an existence, surviving even on the Mars of today? Microbes are tenacious in the extreme when it comes to hanging onto life. On Earth, in the desolate wastes of the Antarctic, some microbes have been calculated to slow down so much that it takes them 10,000 years to metabolise a single molecule of water.

In Antarctica, ice-covered lakes are found in the McMurdo Dry River Valleys, where the average temperature is $-68°$ C (see the illustration on p. 95). Less than four inches of rain falls every year and yet liquid water pockets survive, embedded in insulating ice. In these pockets, communities of bacteria thrive. They are known as 'cynobacteria' and breathe carbon dioxide to live. In fact, cynobacteria are similar to the primitive microbes that polluted Earth with oxygen, their waste product. According to Stephen Giovannoni, Associate Professor of Microbiology at Oregon State University, these bacteria live in conditions that might be very close to those found at the martian polar caps. The north polar cap consists of both water ice and carbon dioxide ice – the two compounds that cynobacteria on Earth need to live.

THE CURIOUS CASE OF METEORITE ALH84001

The belief that life *did* begin on Mars was bolstered during 1996. NASA announced that a meteorite, given the prosaic name ALH84001 (see the illustration on p. 95), contained evidence that might just possibly, at a stretch, suggest that there was once, maybe, bacterial life on Mars (possibly). The less respectable members of the press went crazy with this story: 'LIFE FOUND ON MARS' and similar headlines abounded.

Lake Fryxell in Antarctica. This picture, taken in summer, shows a melted moat surrounding the permanent ice-covered lake. Beneath the ice covering are thriving communities of bacteria. Could the same be true of the martian polar regions? (Photograph courtesy of David Wynn-Williams, British Antarctic Survey.)

Rock Star ALH84001. Discovered in Antarctica on 27 December 1984, this meteorite was carefully stored to prevent terrestrial contamination. Researchers believe that inside the rock are some tell-tale signs that life once existed in its microscopic fractures. What makes it truly amazing is that the meteorite has been positively identified as coming from Mars. (Photograph courtesy NASA.)

Not surprisingly, other scientists, who felt that those headlines were somewhat excessive, tried to put the story into some kind of context and explain that this was not conclusive evidence. As a result, shortly after the original story broke, the headlines changed to 'LIFE ON MARS: WRONG'. Whether or not there really was life in this rock remains a mystery even today – after considerable additional work; but make no mistake – the scientific and press interest in this humble meteorite turned it into a rock star in the truest sense of the words.

The original work on the meteorite was conducted by a group of researchers led by David McKay of NASA's Johnson Space Center. ALH84001 had already been established as having originated on Mars. This much is certain because tiny globules of gas were discovered in the meteorite and the chemical signature of the gas is totally unlike that of Earth's atmosphere. It is, however, identical to the martian atmosphere as measured by the Viking spaceprobes that landed on Mars in the mid-1970s.

The NASA team's attention was drawn by the fact that ALH84001 contained carbonate globules. Carbonate is the name given to the molecule CO_3. It is usually bonded to another atom – most often, a metal. For example, calcium carbonate ($CaCO_3$) makes up the shells of molluscs on Earth and is the principal constituent of chalk, limestone and marble. Since carbonates form when carbon dioxide bonds with water to produce carbonic acid, a study of the carbonates found in the martian rocks promised to reveal information about the flow of water during the red planet's past.

The first oddity appeared when the carbonate globules themselves were studied under an expensive piece of kit known as a 'transmission electron microscope'. Grains of a mineral called magnetite (Fe_3O_4) were found and were shown to be martian in origin. On Earth, magnetite is produced by some bacteria. As the name suggests, it is a magnetic compound and some bacteria manufacture it so that they can use the Earth's magnetic field to distinguish between 'up' and 'down'. As a single discovery, however, this can hardly be thought of as conclusive evidence. The team went further and discovered organic molecules that are known as 'polycyclic aromatic hydrocarbons'. Scientists refer to these as PAHs – to prevent simple conversations about them taking hours to complete. They are hexagonal rings of carbon atoms that can join together in ever increasing complexity. They are readily produced by the decay of life forms on Earth. Although PAHs can be produced by processes other than the decay of dead organisms, the mix of PAHs is different. McKay and his team therefore interpreted them as evidence of decomposed martian life.

The *coup de grace* was the discovery of objects that actually look rather like Earthly bacteria (see the illustration on p. 97). Can appearances be deceptive

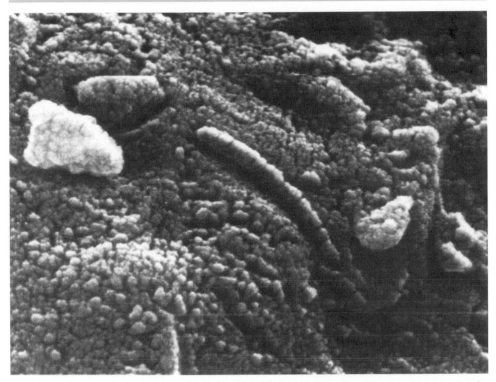

Is this a Martian? An image of the bacteria-like structure found in martian meteorite ALH84001. The jury is still out, deciding if this can be believed or not. Almost certainly, further analysis of the meteorite will not help. Only by going to Mars and prospecting its rocks will scientists be able to determine if this is the fossilised remains of a martian bacterium. (Photograph courtesy NASA.)

though? Despite the superficial similarity of these structures to Earthly bacteria such as *E. coli*, the sizes are very different. The martian structures are 10–100 times shorter than Earthly bacteria. This means that, in terms of the volume, they are between one thousand and one million times smaller. Many have pointed out that this does not leave much space for genetic material. In Chapter 4, I described ribosomes as being those organelles in a cell that build proteins from messenger RNA. The size of a typical ribosome is about 40 nm. The length of the martian structures can reach about 200 nm but are more typically 10 nm. So a ribosome – an essential component of living cells on Earth – could not fit into one of these objects. Some have used this to argue that the structures cannot be the fossilised remains of martian bacteria. I am a little perplexed by this argument because if life evolved on Earth and on Mars independently, then how can anyone be so certain that martian life will need ribosomes to function?

Remember, from Chapter 1, my enquiry directed to extraterrestrials: 'Please explain your biochemistry.' Perhaps any life on Mars was not, or is not, based on DNA. Perhaps the Martians evolved a smaller, compact molecule capable of coding genetic information.

The three individual pieces of evidence presented here do not add up to life because all of them can be attributed to non-biological processes. However, it is the fact that they all appear together, in the same rock, that leads some scientists to believe that the simplest explanation is that they were produced by life.

Whilst it cannot be completely ruled out, terrestrial contamination, after the meteorite was collected, is unlikely. The meteorite collectors of Antarctica are not like errant Indiana Jones figures who simply sling their findings into dirty cloth shoulder bags whilst battling ancient traps/curses/tribes/cults/sects/ Nazis and giant ball-bearings. With more life in the Antarctic being found all the time, however, contamination of the meteorite as it lay in the ice may be much harder to rule out. A recent study of the carbonate globules, by Luann Becker and colleagues (all of the University of Hawaii), suggests that 80% of the organic material is terrestrial in origin. The other 20% closely matches the organic content of meteorites that have nothing to do with Mars. So Becker believes that the 20% portion of the organics in ALH84001 represents what was delivered to Mars during the late bombardment phase.

Meanwhile, the NASA Johnson Space Center team recently provided new evidence of similar tiny bacterial structures in another martian meteorite – but this time, they did so with a quieter fanfare than before. The meteorite that came under their scrutiny this time was the Nakhla meteorite that fell in Egypt. The controversy concerning the real nature of these structures rages on.

Another fascinating aspect of the martian meteorite affair is that it has opened up a biological debate that has been going on right here on Earth for years: namely, that some researchers feel that they have found extremely small organisms in Earth's rocks.

NANOBES AND OTHER TINY LIFE-FORMS

These discoveries all date from the 1990s, with a growing number of geologists and biologists finding structures that look like but are much smaller than, bacteria inside Earthly rock samples. The original discoveries were called 'nannobacteria' and were identified simply by their shape. Their sizes ranged from 20 nm to 150 nm. Normal bacteria range from 150 nm to 50,000 nm. At the same time, tiny organisms only 200–300 nm in size were isolated and grown from foetal bovine serum. These were called (rather confusingly) 'nano-bacteria'.

The most recent contenders, ranging between 20 nm and 1,000 nm in size, were reported by Phillipa Uwins, of the University of Queensland, in 1998. She and her team had been prospecting for a petroleum company when a co-worker remarked that her samples were covered in fungus. She investigated and discovered that although these organisms were fungus-like in appearance they were up to 10 times smaller. Her curiosity piqued, she investigated further, to find that they displayed features that were highly suggestive of a cell membrane with a possible nucleus inside. She stained them with three different DNA tracing chemicals and obtained positive reactions from all three. The most remarkable part of her research is that she then grew these creatures in controlled laboratory conditions – after they had grown quite happily themselves all over her laboratory equipment. She called these nanometre-sized critters 'nanobes', in keeping with the naming convention of microbes, which are measured in micrometres.

Some are still sceptical of Uwins' claims and, in particular, point to the fact that the rock samples were dug up from between 3 km to 5 km below the sea bed off Western Australia. They ask the reasonable question: 'Could we expect nanobes living in those very different conditions to thrive under conditions on the surface?' Although this is a valid question, in all honesty I think it misses the major point, which is: no matter how the nanobes came to be in the rock, they are living and thriving in Uwins' laboratory. Unless she is hideously mistaken, she has identified and grown the *smallest* living organisms ever known. This seems like a most exciting discovery to me because the small size of these creatures virtually rules out replication using the complex organic machinery of known microbes. So, could they be relying on a simpler form of replication that developed first but was overtaken by the more efficient replication of the more complicated microbes? The research continues and the current effort is to extract the nanobian DNA to investigate how much of it there is and what genes it possesses.

I find this all terribly exciting stuff and can see that it certainly helps McKay's case that ALH84001 contains fossil evidence of life. Ultimately, however, the only way to eliminate the objections is to go to Mars and study the rocks *in situ*. Or, at the very least, send a spaceprobe that is capable of returning samples to Earth for study.

PROSPECTING MARS FOR LIFE

The beginning of the third millennium coincides with a vigorous bout of Mars exploration. Within the first five years, robotic rovers similar to Mars Pathfinder's Sojourner will load up capsules on the surface and launch them into parking orbits around the red planet. Then, a French-built spacecraft will head for Mars and perform the equivalent of an interplanetary taxi service by collecting the waiting sample canisters and bringing them back to the eager hands of the Earthbound geologists.

The flotilla of spaceprobes that will journey to Mars in the next decade or two are the advance robotic scouts, preparing the way for the first crewed mission. As has been written before, the effort to get to Mars will mean that it cannot simply be a 'flags and footprints' mission, as some perceived the Apollo Moon landings. Instead, it will be the ultimate field trip, with scientists staying on the planet for months or even years, studying the world and looking for signs of life, either extinct or living.

Where do you land? This is not simply a case of the top brass at NASA standing around a map of Mars with Chief Administrator Daniel Goldin holding a pin whilst his aides blindfold him. In terms of looking for life, the best place to land is anywhere that looks as if it were once covered in water. Perhaps an old lake basin?

Remarkably, scientists have a place on Earth that might provide them with a clue of what to look for. It is called Mono Lake and is situated in California, 360 miles north of Los Angeles. When that venerable American writer Mark Twain visited this place in the 1870s, he wrote: 'Mono Lake lies in a lifeless, hideous desert ... This solemn, silent, sailless sea – this lonely tenant of the loneliest spot on Earth ...' You get the picture. What makes it remarkable is that it is 700,000 years old. It has been created by the run-off of rain-water from the eastern Sierra Nevada. The rain-water has brought with it dissolved salts and minerals that have built up in the lake over time. It is now over twice as salty and 80 times as alkaline as an ocean. Despite its desperate, disheartening desolation (Mark Twain is not the only one who can do altogether amazing alliteration), Mono Lake supports a thriving ecosystem. The water requirements of the distant inhabitants of Los Angeles has led to a lowering of its water level by about 12 feet. This has allowed scientists their first view of an extremophile environment where the water level is dropping.

On Mars, during the Hesperian period when the planet was drying out, it may not have been difficult to find slowly evaporating lakes in the middle of large tracts of desert. Gusev crater on Mars is probably a good example of a structure that could have been an evaporating lake (see the illustration on p. 101). What geologists would like to identify are mineral structures called tufas. They are made of a type of limestone that forms from the precipitation of the minerals in the lake. They are usually found under the water but in the case of a dried-up lake on Mars they would obviously be visible as a surface feature. The reason for the interest in them is that, on Earth, they are great places to find fossils. If life on Mars existed in such lakes, then their fossilised remains might be expected to be found in martian tufas.

An evaporated crater on Mars? Gusev crater is shown here on images taken by one of the Viking spacecraft. It is thought to have been supplied with water by the Ma'adim Valley, billions of years ago. If this is the case it will be one of the best places to look for martian fossils. (Photograph courtesy NASA.)

Jack Farmer, of Arizona State University, has been studying the tufas of Mono Lake in order to know what to look for on Mars. In 2001, Mars Surveyor 2001 Orbiter will carry an instrument capable of pinpointing specific chemical signatures to within 100 m. This will allow tufas to be identified and, shortly into the new millennium, mission planners will be able to start drawing up a short-list of possible crewed landing sites. The robotic landers may also send back samples that warrant further, human investigation.

Mars Express is Europe's first sole mission to the red planet and launches in 2003. The main scientific focus of Mars Express is to find and study the remnants of martian water. To this end, the spacecraft will carry eight instruments, one of which will land on the planet.

The Mars Express lander is called Beagle 2 – after the *Beagle*, the ship in which Charles Darwin sailed – and is an entirely British endeavour led by Colin Pillinger of the Open University. Beagle 2 will detach from Mars Express five days before rendezvous with the planet. It will hit the martian atmosphere and be slowed by friction until parachutes can be opened to slow it even more. Rather like NASA's hugely successful Mars Pathfinder mission, Beagle 2 will inflate large protective gas bags around itself, eject the parachutes and drop the last kilometre to the surface, finally bouncing to rest. This strategy prevents the parachutes from descending onto the lander and smothering the cameras and instruments.

Once on the surface, Beagle 2 will look for the evidence of life, either past or present. A microscope will be used to examine soil samples in unprecedented detail. A robotic arm will collect rock fragments and soil so that they can be analysed for the presence of organic matter, water or minerals that have been deposited in the past by running water. The lander will also possess an ingenious device, dubbed 'the mole'. This will be capable of crawling short distances across the martian surface and burying under rock to analyse the more protected soil found there.

WHERE ELSE SHOULD WE LOOK?

The other world name that rolls off the tongues of exobiologists is that of Europa. In the previous chapter I introduced this intriguing little world by discussing the tidal friction that many believe keeps its subsurface liquid. Scientists in Antarctica again see parallels between the conditions in some places on Earth where they find life and those that might prevail on Europa.

A Russian research outpost, Vostok Station, has been uncovering examples of micro-organisms from deep in the ice, using a drilling rig developed at the St Petersburg Mining Institute in 1974. For about 20 years, principal scientist S.S. Abyzov, of the Russian Academy of Sciences, has been reviving microbes from the ice samples. It is estimated that some have lain dormant for 400,000 years, buried in ice a few kilometres thick. Another amazing discovery came from Vostok Station in 1996, when the Russian Academy of Sciences announced that a large lake of liquid water had been discovered, by radar, lying beneath 3.7 km of ice. They have currently drilled down to within about 100 metres of it but, wisely, are holding back from actually breaching the lake until they can be certain of studying it without causing contamination.

Lake Vostok, as it has been called, is about the size of Lake Ontario (48 km × 224 km × 484 km) and recent radar work suggests that about 50 metres of sediments are lying on the bottom of the lake. The overlying ice has been analysed in collaboration with Richard Hoover from NASA's Marshall Space Flight Center and many microbes that have never been seen before have been discovered. To aid the work of classification, the scientists have given them nicknames such as 'Mickey Mouse', 'Klingon', 'porpoise', 'sphere' and 'left-over turkey', which leads me to suspect that the Americans in the project are responsible for the naming; had it been the Russians, the names might have been 'empty vodka bottle', 'furry hat', 'the Kremlin' ...

The scientific anticipation of what might be found in Lake Vostok is difficult to overstate. No one knows why it should still be liquid, or how it got where it is – or anything really! Analysis of the ice layers closest to the lake indicate that it has probably been cut off from the rest of the world for between 500,000 and one million years. If it contains a thriving community of extremophiles this would surely increase the chances of finding life in the subsurface oceans of Europa.

Speculative missions are already being drawn up to map the thickness of the Europan ice crust. One such mission is the Europa Orbiter, which is being designed by a consortium of scientists led by NASA's Jet Propulsion Laboratory. The first mission objective is to test beyond doubt that there is a global ocean under the Europan ice sheets. As the name implies, this spaceprobe will orbit the moon, rather than fly past as the Galileo and Voyager probes did. Using radar, Europa Orbiter will be able to detect the presence of liquid water because radar responds differently to water than it does to ice. This will allow it to simultaneously begin the second mission objective of mapping the three-dimensional structure of the moon. Cameras will take unprecedented images of the icy surface, looking for areas of recent or even current activity where water is welling up from below. Together the photographs and the radar data will be used to select potential landing sites for a much more ambitious mission to this fascinating place.

It is hoped that a landing vehicle will settle on the thinnest part of Europa's ice crust and literally melt its way down into the ocean. At present a team of scientists from Cornell University are testing a probe that can melt its way through thick ice. They call their hypothetical Europan mission, Odysseus.

Once a probe has successfully melted through the ice and penetrated the water, mission requirements call for the release of special underwater rovers, called 'hydrobots', that could cruise the global ocean of Europa looking for signs of life. Wherever scientists find water on Earth, they invariably find life. Could the same be true for the rest of the Universe? The Europa missions will find out.

In Chapter 1 I stated that I would stick as closely as I could to science, rather than flights of fancy, in my consideration of alien life. With that as a stated aim, I do not want to descend into the ideas about life in the atmosphere of Jupiter or on the Sun-scorched wastes of Mercury but I will just mention Venus. Although the hellish conditions of Venus are thought to be completely unsuitable for life now, in the remote past Venus, too, was probably as clement as the Earth. By the same reasoning that led us to believe that Mars might have fostered the emergence of life, Venus may also have done so.

The increase in the Sun's luminosity had a devastating effect on Venus, driving it towards the runaway greenhouse effect that scorches that world today. Even if life began there, it seems inconceivable that current technology could find its remains. It may even be the case that the prevalent conditions of Venus have already destroyed any evidence of past life.

Despite this chapter's musings, one thing is certain; there are no other worlds in the Solar System with a widespread surface-dwelling ecosystem similar to the one we have on Earth. If we are to find other thriving biospheres, we will have to look for planets around other stars. Although this concept might seem like science fiction, I will explain in the next chapter that the technology required to search for Earth-like planets around other stars and to determine whether they are inhabited, is within humankind's grasp.

Chapter Eight

Searching for inhabited

---

planets

The fact that many respected scientists now believe life, in some form, may be present on several bodies within the Solar System cannot help but lead us to think that life on a galactic scale is widespread. In 1982, long before the current frenzy of interest, the International Astronomical Union set up its 51st Commission and called it 'Bioastronomy, the search for extraterrestrial life'. The Commission's job is to help organise a coherent investigation into the search for extraterrestrial life and it has drawn up seven main principles:

> To search for planets around other stars.
> To study the evolution of planets and their possibilities for life.
> To detect extraterrestrial radio signals.
> To detect organic molecules in the Universe.
> To detect primitive biological activity.
> To search for signs of advanced civilisations.
> To collaborate with other international organisations, such as those
> devoted to biology, astronautics and so on.

Decades before, a young American astronomer, Frank Drake, had pondered the existence of life in the rest of the Milky Way. In the early 1960s he was to meet with 12 other scientists – at that time almost the entire world contingent of astronomers who were interested in actively searching for intelligent extraterrestrials. As a way to organise his thoughts, he wrote down the factors that he felt would determine the number of intelligent civilisations. They can be summarised as the following:

> The formation rate of suitable stars in our Galaxy.
> The fraction of these stars that possess planets.
> The average number of potentially habitable planets per star.
> The fraction of potentially habitable planets on which life develops.
> The fraction of inhabited planets on which life develops intelligence.
> The fraction of intelligent civilisations that develop radio technology.
> The average time that a technologically advanced civilisation can live.

The definitions themselves are somewhat fluid. For example, the formation rate of suitable stars can be separated into the formation rate of stars in the Galaxy and the fraction of those stars that are suitable to possessing habitable planets. This does not really matter; what *is* important is that, as Drake intended, the considerations provide us with a way to organise our thoughts and breakdown the problem into more manageable chunks.

You will realise that I have already discussed some of these considerations during the course of the book. For instance, I have advocated a viewpoint that life will develop on any habitable world; so we can drop consideration 3 entirely and simply restate consideration 4 so that it directly refers to consideration 2: 'The fraction of those planets on which life develops'. The formation rate of stars is something I described in Chapter 3, and in Chapter 5 I touched upon the way in which planets form when I discussed the formation of the Earth. What I did not refer to is precisely how common we expect planets to be throughout the Galaxy. Nor have I even begun to think about the rise of intelligence on Earth or elsewhere. A good job, too, otherwise I would not have anything left to talk about in this chapter and the next one.

PLANETS BEYOND THE SOLAR SYSTEM

1995 was a monumental year in the search for planets orbiting other stars – so-called 'extrasolar' planets (from the Latin 'extra', meaning outside or beyond). After years of false alarms and misidentifications, the first confirmed planet orbiting another star was discovered. It was a shock for two reasons. Firstly, the astronomers did not actually see the planet; they inferred its existence from observations of the star. Secondly, the new discovery could be loosely described as 'the wrong kind of planet in the wrong type of orbit'.

To understand the first surprise, imagine trying to see a table tennis ball held next to a searchlight. If the searchlight were to be turned off, it would be no problem for you to see the two distinct items. If the searchlight were to be turned on again and pointed towards you, you would be so dazzled by the light, that seeing the table tennis ball through the glare would be impossible. This is the situation in which astronomers find themselves when looking for planets. Stars are incredibly large and bright, whilst planets are incredibly small and dim by comparison. The diameter of the Sun is approximately ten times greater than the diameter of Jupiter, which, in turn, is about ten times greater than the diameter of the Earth. Planets emit no light of their own; instead, they simply reflect starlight. This means that they will always be much dimmer than stars.

The vast majority of us are aware that the planets of our Solar System are in orbit around the Sun; but this is actually a slight oversimplification. To be precise: any two bodies are in orbit around a fixed point between them, known as the 'barycentre'. The exact position of the barycentre is determined by the relative strengths of the gravitational fields between the two objects. The gravitational field of an object depends upon the amount of matter it contains. A massive object such as a star creates a stronger gravitational field than a planet creates.

Remember classroom science, when your teacher tried to teach you how to balance objects. Using nothing more than a length of wood (usually a ruler) and a triangular block (technically called a fulcrum) to place it on, you could determine how to balance objects of various sizes. You did this by sliding the triangle towards the heavier object (see the illustration below). In each case you were, in fact, finding the barycentre of the system.

If two stars of equal mass are in orbit around each other, then the barycentre is halfway between them. If we look at another star system, in which one star is twice the mass of the other, then the barycentre is located a quarter of the way from the larger star to the smaller star. As the mass

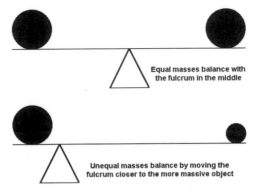

Balancing objects. Equal objects balance when the fulcrum is halfway between them. The larger the difference in the masses of the balancing objects, the closer to the heavier one the fulcrum needs to be.

difference becomes bigger and bigger, so the barycentre will be closer to the more massive object. In the case of the Sun and the Earth, the barycentre is actually inside the Sun. This is not the case for Jupiter. The barycentre between the Sun and Jupiter is just outside the surface of the Sun. As Jupiter orbits this point once every 12 or so years, so must the Sun. The Sun performs a little pirouette, taking about 12 years to complete it.

Although Jupiter has the largest effect, all the other planets also contribute and pull the Sun this way and that. Viewed from above, the Sun follows a path that looks somewhat like a drunkard attempting to get off a merry-go-round (see the illustration below).

In principle, this swirling motion could be detected by measuring the star's precise position over the course of several years. In practice, the blurring effects of our atmosphere render the task almost impossible from the Earth. This will not always be the case as more and more sophisticated systems are developed that compensate for the atmospheric effects on image quality.

However, the first planets were discovered by exploiting the motion of a star around its barycentre in another way. We are all familiar with the way in which a ringing bell changes pitch as we pass it in a car or on a train. Another familiar example is the way emergency sirens on police cars, fire engines and ambulances change as they pass us. In these cases the change in the sound is caused by the Doppler effect.

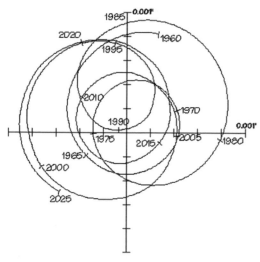

The Sun's path around the Solar System's barycentre. The path of the Sun is shown for the period 1960 to 2025. The movement is caused by the interplay of gravitational forces between the Sun and the nine planets. Because the nine planets' orbits are all different, the combined force on the Sun is constantly changing. (Photograph reproduced courtesy of NASA.)

In 1845, Christiaan Doppler stated that when there is motion between a source of sound and the observer, the pitch (frequency) will change. If the distance between the source and the observer is decreasing, the frequency will rise as the sound waves are squashed. Motion away from each other would cause the sound waves to stretch and the pitch to drop.

The Doppler effect works equally well with light as it does with sound. In the case of light, blue light has a higher frequency, therefore a shorter wavelength, than red light. So a decreasing motion, between source and observer, decreases the wavelength of light, causing it to be blueshifted. On the other hand, an increasing distance between the source and the observer stretches the wavelength of light, creating a redshift.

It can therefore be imagined that the motion of a star around its barycentre will sometimes cause it to be travelling towards the Earth and sometimes be travelling away from the Earth. So, any observations of the light emitted by a star being pulled by a planet will oscillate between redshifts and blueshifts. Identifying this motion for any particular star is made easier by the fact that gases surrounding the star will absorb certain specific wavelengths of light. When the starlight is passed through a prism and split into a spectrum, the wavelengths that have been absorbed will show up as single black lines in the rainbow of colours that make up the spectrum. In reality, there are usually hundreds of these absorption lines throughout the spectrum. Each chemical element produces its own unique pattern. In effect, they are providing a chemical fingerprint, as I described in Chapter 2.

By identifying these patterns, astronomers can precisely measure the wavelengths of the lines, determine which chemical element they originate from and whether they have been redshifted or blueshifted. By repeating these observations regularly over many years, any orbital motion of the star around a barycentre will be determined. Theoretical predictions state that the orbital velocity of the Sun caused by Jupiter is just 12.5 m s^{-1}. If this seems slow, the Earth's gravitational pull moves the Sun by only 0.09 m s^{-1}.

The detection of Earth-like planets is inconceivable with the Doppler effect method because the boiling surfaces of stars are expected to be moving faster than this and will simply smear the orbital motion into undetectability. Jupiter-like planets, however, are within the bounds of detectability due to the amazing precision of a new generation of spectroscopes. Using these sensitive devices, astronomers can pin down stellar velocities to within 3 m s^{-1}.

On 6 October 1995, Swiss astronomer Michel Mayor announced that he and Didier Queloz, of Geneva Observatory, had detected the signature of a planet around a star called 51 Pegasi. The planet was calculated to be about half the mass of Jupiter. So far, so good. Instead of taking a dozen years to perform a single orbit, as Jupiter does, this new world took just 4.2 days. We think dog-years are bad, with every one of our years said to be equal to about seven of theirs but on 51 Pegasi b a year lasts only 4.2 days, giving them about 87 years to our one year!

This discovery sent astronomers into equal mixtures of rapture and panic. The identification of the first planet outside our own Solar System was a momentous step forward but the planet was in the wrong orbit. Taking just 4.2 days to orbit its star meant that the planet had to be in an orbit with a radius of just 0.05 AU. Even Mercury – the planet closest to our Sun – has an orbital radius eight times larger than that of 51 Pegasi b.

Remember that in Chapter 5 I confidently stated that the conditions near to the Sun would permit the formation of only rocky planets. Well, even the largest rocky planet, the Earth, is still 333 times less massive than Jupiter and so the most reasonable explanation is that 51 Pegasi b is a gas-giant planet like Jupiter, that formed far away from the star, beyond 5 AU, then migrated inwards, stopping just in time before it plunged to a premature and fiery death inside 51 Pegasi.

Current research suggests that this could indeed be possible if the swirling disc, from which 51 Pegasi b condensed, contained a lot of material. As the newly-formed planet ploughed its way around this disc, the friction robbed it of orbital energy and it spiralled inwards, ending up in its current orbit. As fantastically improbable as this sounds, other astronomers – notably Geoffrey Marcy, of San Francisco State University – who were also running planet search programmes, began digging through their data and they too found previously undreamed of planetary signatures: 51 Pegasi b was not a freak.

To date, about a dozen-and-a-half extrasolar planets have been discovered, using the Doppler technique. The vast majority of stars surveyed, however, have shown nothing detectable. As a rough estimate, the current data suggest that about 5% of stars like the Sun, in the solar neighbourhood, possess 51 Pegasi b-like planets. The other 95% may have no planets at all, or they may have systems similar to ours (which will take longer to detect). As yet, no one knows. The Doppler searches will continue and over the next decade are destined to provide us with the majority of our knowledge of planets outside the Solar System. Space missions, using other methods of detection, have been proposed by both European and American astronomers. They sound like flights of fancy, yet are expected to be launched early in the second decade of the new millennium and will allow family portraits of extrasolar planetary systems to be taken.

PLANET SNAP-SHOTS

Taking pictures of planets is very difficult. Photographers everywhere know that persuading everyone to smile in a group photograph is difficult enough – but imagine trying to take a group photograph when some members of the family are so badly illuminated that they might not even appear on the finished print.

Planets are, unfortunately, dim objects. They emit no visible light of their own and are, typically, one billion times fainter than the stars they orbit. The problem is slightly alleviated by observing in the infrared, where planets *do* emit radiation in the form of the heat they have picked up from their central star during the course of the day. Stars like the Sun emit less infrared than visible light and so the contrast between a star and a typical planet is only(!) one million to one at infrared wavelengths beyond 10 μm. These wavelengths are impossible to observe from the surface of the Earth because molecules of gas in our atmosphere absorb them. Hence, a telescope to detect planets at infrared wavelengths would need to be launched into space.

As if this was not making the mission hard enough, a single telescope will simply be insufficient. Instead, a small armada of four or five identical telescopes will have to fly in precise formation, combining their observations in such a way that the infrared emission from the central star is almost completely cut out, leaving only the planets to shine forth and be captured with special electronic cameras known as 'charge-coupled devices' (CCDs).

A NASA committee investigated this amazing possibility and concluded that such an observatory would be feasible within 20 years. It would be able to detect Earth-like planets around the nearest 1,000 stars. Their report did not stop there; it went on to explain how scientists from around the world had devised the means for a straightforward test to determine if those planets were also inhabited. This is the scientific equivalent of not just finding the dartboard but throwing a bull's-eye at the same time.

The method relies on splitting the planet's infrared light into a spectrum and analysing the absorption lines. These lines will have been formed by the atmosphere of the planet under scrutiny. The ground of that planet will radiate at many infrared wavelengths but on its way out of the atmosphere the planet's atmospheric gases will superimpose their chemical fingerprints by absorbing certain wavelengths of the radiation.

As early as 1970, James Lovelock wrote about detecting life on alien worlds by analysing the gases found in their atmospheres. Consider the Earth. I have explained that the initial oxygen build-up in the atmosphere was caused by the metabolism of microbacteria. Life cannot help but have a profound effect on the atmosphere of its home planet. The question is: can we learn to recognise atmospheric combinations that are produced by life? Not forgetting, of course, that life elsewhere may be very different from us.

The spectral signature of carbon dioxide would be a good indicator that the planet in question possesses an atmosphere. Of the five largest worlds in our inner Solar System (Mercury, Venus, Earth, Mars and the Moon), three have atmospheres (Venus, Earth and Mars) and all possess significant proportions of carbon dioxide in their atmospheres. Only Earth possesses significant amounts of water and ozone.

On Earth, water makes the planet habitable and ozone is made of oxygen, the by-product of plant life and certain microbial metabolisms. If all life were to come to an end tomorrow, the amount of oxygen would gradually decrease, as this highly reactive element would find other elements to join with. This happened on Mars when the oxygen, released by the break-up of water molecules, reacted with the surface rocks.

The detection of water and ozone would not constitute absolute proof, however. There are some non-biological reactions which could be imagined that would create large amounts of oxygen. For example, a greenhouse effect could temporarily increase the amount of oxygen in an atmosphere by promoting the break up of water into hydrogen and oxygen in the high atmosphere.

A better subject for a search would be atmospheres that contain pairs of molecules known as 'redox couples'. In a redox reaction, one atom or molecule donates an electron to any atom or chemical that appears to need it. Oxygen is incredibly good at pinching electrons from certain other chemicals and so any chemical that donates its electron is said to be 'oxidised', regardless of whether or not oxygen was actually involved. This convention is rather like all vacuum cleaners being termed 'hoovers', when, in fact, Hoover is a brand-name of a vacuum cleaner. Any chemical that accepts an electron has its electrical charge lowered by 1 because an electron carries an electrical charge of -1. As a result, the chemical is said to have been reduced.

A chemical reaction in which two chemicals join because one donates an electron and the other accepts it, is known as redox reaction. Water is produced by a redox reaction between hydrogen and oxygen. The oxygen is reduced by the electron it takes from the hydrogen; thus, the hydrogen has been oxidised. Left to their own devices in an atmosphere, redox couples would all combine. So, if an atmosphere can be found with redox couples living independently side by side, this would be a strong implication that some form of process is taking place that constantly replaces the individual components.

The detection of a redox couple in the atmosphere of an extrasolar planet would cause a flurry of activity in Earthly chemistry laboratories to see whether non-biological processes could cause the excess. If not, the chances are very high that what has been detected is none other than the living breath of alien life-forms. In the case of the Earth, the most detectable redox couple caused by the action of life-forms is that of ozone and methane.

The successful detection of an infrared spectrum produced by an Earth-like planet is no easy task. It will take about a month of dedicated non-stop observation to collect enough of the feeble radiation to be able to split it into a spectrum and search for life's molecules. The longest-ever exposures by a telescope so far are the 10-day exposures required to capture the Hubble Deep Fields, which I referred to in Chapter 3.

The NASA committee which investigated the redox detection produced a document called the Roadmap for the Exploration of Neighbouring Planetary Systems (ExNPS – pronounced 'x-nips'). It culminated in the proposal to build the Terrestrial Planet Finder – a collection of four 3.5-metre telescopes, each of which would fly in its own little spacecraft. These would then fly in precise formation, spread over a distance of one kilometre, using laser beams to keep them perfectly aligned. The images from the four telescopes would be beamed to a fifth spacecraft, which would combine the images and then transmit them to eager astronomers on the Earth.

The reason for using several telescopes in parallel is that they can simulate a single telescope with a much larger mirror than the individual components possess. Also, the beams of light from the component telescopes can be combined in such a way that they reduce the overwhelming brightness of the central star. This is an impossible task for a single telescope. In the case of the Terrestrial Planet Finder, it is estimated that the central star could be dimmed by as much as 100,000 times. This would make any planets around these stars easily visible. The concept is known as a 'nulling interferometer'. Current plans for this exciting space mission call for a launch date in 2011 (see the illustration on p. 114).

In Europe, a consortium of astronomers have investigated very similar lines of reasoning and have put together the Darwin project, which the European Space Agency could choose to launch early in the 2010s. Both missions have as their primary goal the detection of life's signatures on Earth-like planets around other stars.

Astronomers had a 'dry run' at detecting life on a planet in 1990. The Galileo spaceprobe flew past the Earth on its circuitous route to Jupiter and astronomers used its instruments to survey Earth just as they would any other planet. The question they were really interested in was: could Galileo detect life on the Earth?

Famous astronomer and advocate of extraterrestrial life, Carl Sagan, investigated this question and found that the answer was 'yes', on three major counts. The first was the detection of molecular oxygen and methane – one of

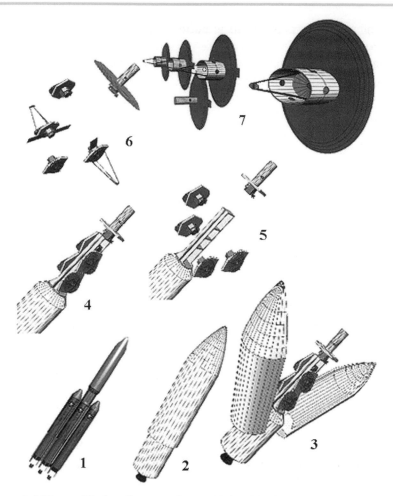

The Terrestrial Planet Finder. Current plans call for this flotilla of space telescopes to be launched in 2011. Here the seven stages from launch to final configuration are shown schematically. (Illustration reproduced courtesy of NASA.)

the redox couples. In addition to this, the Earth strongly absorbed red light, especially over the continents. This behaviour was governed by the presence of chlorophyll. The molecule is essential for both plant-life and certain microbes because it allows photosynthesis to take place. This is one of the principal ways in which a plant obtains the energy it needs to live. With the absorption of a photon of sunlight, plants can turn carbon dioxide into carbohydrates, which are energy packets that can be used to power the growth of the plant. Significantly, the Galileo spaceprobe also detected radio transmissions from Earth that were clearly not of natural origin. This principle of eavesdropping on stray radio signals is the strategy of traditional SETI astronomers.

Two years later, the Galileo probe again swung by the Earth. This time, the science team used the instruments to study the Moon. It saw nothing that could indicate life: no redox couples, no chlorophyll, no radio transmissions – just a dead world. The results of this work provided one of the strongest inspirations for the scientists who are now working on the Darwin and Terrestrial Planet Finder missions.

HOW MANY PLANETS ARE THERE?

At the moment, the number of stars that possess planets remains unknown. There are about 200 billion stars in our Galaxy alone. If the Sun is typical and if each star possesses an average of nine planets, that is almost 2 trillion planets – more noughts than even Leonardo DiCaprio has on his pay cheques.

The extrasolar planets found so far give reason to hope that planet formation is simply a by-product of star formation. As I mentioned in Chapter 5, when I discussed the formation of the Earth, the more astronomers look at young stars, the more they find them associated with dusty discs that appear conducive to planet formation (see the illustration on p. 116 and also look again at the illustration on p. 67).

It had previously been assumed that the orbit of a planet in a binary system would not be stable, as the planet would be orbiting a star which itself was orbiting another star. Since binary and multiple star systems account for around 50% of stars in our Galaxy, the assumption was that the number of stars that could possess planets was automatically half of the total number of stars and probably much less than that when a whole load of other mitigating factors were also taken in account.

That viewpoint, however, has now changed because 16 Cygni B is one half of a binary; yet Doppler observations have shown that it is orbited by a planet fully 1.5 times more massive than Jupiter, following an elliptical orbit within average radius of about 1.6 AU. Theoretical calculations also show that close binary stars might have planets in orbit around both of the components. Indeed, the young star GG Tauri possesses a dust disc that surrounds its two component stars and there are also small-scale dust discs around the individual components. If this were not enough, the whole system itself is in orbit with a second pair of stars, so it is a quadruple star system. There are still some orbital configurations in binaries that make planetary systems unlikely but on the whole it looks as though Tatooine – the desert planet in *Star Wars*, that orbits two stars – could, after all, exist.

This Hubble Space Telescope image clearly shows a dusty disc of material surrounding AB Aurigae. The cross shape is caused by the instrument necessary to block out the light from the star so that the fainter disc can be clearly seen. (Photograph reproduced courtesy of C.A. Grady (National Optical Astronomy Observatories, NASA Goddard Space Flight Center), B. Woodgate (NASA Goddard Space Flight Center), F. Bruhweiler and A. Boggess (Catholic University of America), P. Plait and D. Lindler (ACC, Inc., Goddard Space Flight Center), M. Clampin (Space Telescope Science Institute), and NASA.)

So, I shall be optimistic about the possibility of planets around stars and say that in many cases they *do* form. Personally, I think the proportion is at least 50% and perhaps a lot more. I predict that there will be plenty of sites for life in the form of planets throughout our Galaxy. There are certainly enough astronomers in the world currently dedicated to finding these planets and I anticipate an unending stream of discoveries every year from now on.

Chapter Nine

ET IQ

I n the previous chapter I described the way in which bioastronomers can search for the evidence of life on planets that are many light years away. Microbes are one thing; intelligent beings are another – as I am sure anyone who has tried to have a conversation with a petri dish of penicillin will agree. The brutal fact is that just because life has developed on a planet, it does not automatically follow that intelligence will develop there.

Evolution does not seem to have a kind of predetermined route to intelligence. If it did, the dinosaurs would be the ones driving around in motor cars, living in high-rise apartments and worrying about the stock market. Also, by necessity, their motor-car suspension technology would be far in advance of ours.

The questions we need to ask and answer in this chapter are 'What is intelligence?', 'How did it evolve?' and 'What are the chances of it evolving elsewhere?' So let's get started.

WHAT IS INTELLIGENCE?

The *Concise Oxford Dictionary* defines 'intelligent' as: '1. having or showing intelligence: 2. quick of mind; clever', neither of which is particularly helpful in a scientific context. By the time we get to the third definition, things are looking up: '3. able to vary its behaviour in response to varying situations and requirements and past experience.' Ironically, this definition is stated to apply to devices or machines.

The fact is that intelligence – like life – is difficult to define. Many have tried to quantify it in the form of intelligence tests and the infamous Intelligence Quotient (IQ) springs to mind. Francis Galton pursued this line of research and between 1884 and 1890 he ran experiments at the Science Museum, South Kensington, where he offered to test people's intelligence – for a modest fee, of course.

Today many psychologists believe that a person is born with an innate intelligence, or, perhaps more precisely, an innate potential intelligence. This level can be deduced by subjecting the individual to an IQ test consisting of puzzles such as: 'If dog is to puppy, cat is to (foal, kitten, tyrannosaurus rex).' You simply choose the answer you think is right and move on to the next question. There will also be mathematical and spatial relationships to exercise your brain. At the end of the test, you are told by your examiners whether you are clever (or not clever). Yet, from my humble vantage point, it seems that these tests often rely more on testing the individual for their knowledge and ability to work under stress.

IQ tests obviously test something but whether that something is truly intelligence, or simply the ability to take tests, I am not sure. I suspect that the truth lies somewhere in between. I can already hear the letters from outraged psychologists landing on my door-mat, so let me present an extreme example to illustrate my point. An alien lands tomorrow and is whisked away to the psychology department of a local university. If the psychologists simply sit this little green man/woman/thing down in front of an IQ test, he/she/it will almost certainly score zero. After all, how is he/she/it supposed to know what a cat is? The fact that their spaceship is sitting on the White House lawn, however, tells you this may not be the best way to determine whether or not this particular individual is intelligent. Context is everything. As with all debates in all sciences – including those in astronomy – psychologists appear entrenched in two camps: pro-IQ and anti-IQ. A consequence of this is that there is a continuing debate about whether it is fair to assign the same IQ test to different ethnic communities; so in terms of testing alien intelligences, IQ tests are obsolete.

I think that the biggest problem with intelligence testing is that the test itself relies on right and wrong answers. If you think about it a little, 'right' and 'wrong' are simply two opposite points with a much larger grey area between. Many questions which require the application of intelligence simply do not have right and wrong answers. This is true even in the mathematical and logical arena. If one of my students works through a lengthy calculation but makes a mistake at the end, I would still give that person the majority of the marks.

Despite the widespread popularisation of IQ and its associated number 'g' – the general intelligence factor – some psychologists are becoming increasingly vocal about their alternative viewpoints. One alternative is the multiple intelligences approach. For much of the twentieth century, the idea of multiple intelligences has been quietly refined. Now Howard Gardner of Harvard University, has proposed at least eight (possibly nine) separate intelligences that we all possess to a greater or lesser degree. I present them here in no particular order:

Linguistic Intelligence. The ability to talk, write and use language.

Logical–Mathematical Intelligence. The ability to deal with abstractions and concepts that may be far removed from everyday experience. This is the intelligence most closely tested in IQ tests.

Musical Intelligence. The ability to create and perform music. It encompasses melody, harmony and rhythm.

Spatial Intelligence. The ability to perceive the world accurately and transform it at will into paintings and sculptures.

Kinaesthetic Intelligence. The ability of bodily control. Dancers and athletes possess this ability in bucket-loads, as do a lot of actors.

Intrapersonal Intelligence. The ability to sense moods and feelings in oneself and to use this as a guide for our behaviour.

Interpersonal Intelligence. This ability is similar to intrapersonal intelligence, except that it applies to one's ability to sense moods and feelings in others. It is, therefore, a pretty good intelligence to develop if you want to be a psychiatrist. It would even be useful for politicians so that they can truly represent their country's wishes.

Naturalist Intelligence. This ability makes it possible for people to recognise and categorise natural objects. Charles Darwin had it in spades.

Gardner is also considering the evidence for a ninth form of intelligence, called Existential Intelligence. This would concern the ability to capture and ponder the big questions of existence: life, death, the Universe – that sort of stuff.

With the proviso that I am an astronomer (so what do I know about psychology anyway?) I think that the 'multiple intelligences' approach is the way to view our intelligence capabilities. I find it unacceptable that a scientist might be held up automatically as more intelligent than a linguist just because the chances are that the scientist is better at solving the logical associations presented in IQ tests. It seems to me to be the same as trying to compare apples with oranges. They are both fruit but one can hardly be said to be a better fruit than the other. So it should be with intelligence.

Which fruit would you link to which intelligence? Now there is a psychology project ...

These musings about multiple intelligences leads me to be a stubborn physicist again and to apply my reductionist thinking to wonder whether there is an underlying principle that is the real intelligence factor; some common ground shared by all of these manifestations of intelligence. Thankfully, other, cleverer people have wondered the same thing before and have decided that the root of intelligence lies in our ability to deal with novel situations. Another way to think of this is to consider our problem-solving ability, which often requires creativity and flexibility in our thinking.

Whenever we are presented with a new problem to solve – be it 'I have no food in the house and twelve dinner guests arriving in half an hour, so what shall I do?', or 'How do I harmonise with a C diminished 5th chord followed by an A major 7th suspended 2nd while Fred whistles *Yankee Doodle Dandy*?', or, indeed, 'Is there life on other worlds?' – in order to deal with it, the first thing we would do is to formulate a plan of action. This is one of the roots of intelligence: the ability to plan. Whenever you are presented with a problem, you conceive a plan and take in as many contingencies as possible to make your plan as foolproof as possible.

Storytellers play on our intelligence in thrilling books and movies about people who are engaged in dangerous missions. We are impressed by the coolness with which our lead character handles impossible situations because he/she has thoroughly planned for all contingencies. Then, just when you think the outcome is a foregone conclusion, something totally unexpected occurs and the hero is forced to abandon his/her intelligence and rely on those primitive, animal-level instincts. In other words, he/she runs like hell. The opposite is also very popular, in that, near the start of the story everything goes hideously wrong because our hero has not planned well enough. The rest of the movie is an exercise in higher intelligence as he/she tries to figure a way out of the mess.

THE HUMAN BRAIN

At least everyone agrees that the brain is the seat of intelligence. A human brain contains roughly 100 billion nerve cells. That is about half the number of stars in our Galaxy. The brain is situated at the upper end of the spinal column and forms the 'front end' of the central nervous system.

The spinal cord is joined to the brain by the *medulla oblongata*. This is the location of the control centres for respiration, heart-beat and blood pressure. Overlying this, at the base of the skull, is the *cerebellum*. This part of the brain oversees complicated muscular movements. Over this, the *cerebrum* is the part of the brain that is divided into two hemispheres. It is where sensory input is processed. The surface of the cerebrum is known as the 'cerebral cortex'. It is a 2 mm-thick collection of cells (called grey matter) that spawn fibrous connections (called white matter) to all other parts of the cortex and the central nervous system. Importantly, in the context of this chapter, medical imaging devices that can detect brain activity have shown that intelligent behaviour (by which I mean novel-problem solving) is associated with the cerebral cortex.

A comparison of the brains in vertebrate animals – those with backbones – is very interesting. The earliest vertebrates, indicated by the fossil record, have the smallest cerebra. In fact, in all mammals the cerebrum is the largest part of the brain and looks like a wrinkled walnut. This wrinkling increases surface area in a restricted volume. According to William Calvin, of the University of Washington, Seattle, if you were to flatten out the cerebral cortex of a human it would cover an area equivalent to four sheets of typing paper. But who would want to write a letter on it?

The human brain is not the largest on the planet, however. Whales have colossal brains, yet we are the ones who study *them*, not *vice versa*. So to quote an old phrase in a new context, 'size doesn't matter; it is what you do with what you've got that counts'.

If sheer size does not matter – or, at least is not the beginning and ending of intelligence – then what *does* matter? Some scientists approach this problem by trying to figure out how intelligent other animals are.

IF I COULD TALK TO THE ANIMALS ...

Nowadays there is an increasing scientific effort to communicate with animals. This is because there is one fundamental difference between humans and animals: only humans possess language. Language is the way in which we communicate our intelligence. It may even be the vehicle of intelligence. After all, what good is having a Nobel prize-winning idea if you cannot tell anyone about it.

In most of us, the ability to process language and talk has been traced to an area of the brain located just above the left ear. Our closest genetic cousins, the chimpanzees, lack this left lateral language area and, as a result, appear to be incapable of language.

'But all animals make noises to communicate!' I hear you cry – and you are correct. But the sounds are not language. They are sounds which convey emotional states of mind, such as pleasure, unhappiness, anger and so on. Their utterance is associated with activity in the brain located near the *corpus callosum* – a band of fibres connecting the two hemispheres of the brain. This animal verbalisation of emotions goes on in humans too. If you are suddenly scared you will scream, rather than say 'Golly gosh, that frightened me'. Your scream is your corpus callosum speaking, not your left lateral language area.

In the animal kingdom, it is strictly: one sound, one emotion. Animals do not string sounds together to give them new meaning. As an illustration of this idea, rather than as a rigid scientific example, 'whee' is an exclamation of exhilaration that you might make when riding a roller coaster and 'eek' is an exclamation of surprise. Put them together (while being a little slovenly with the 'h') and, in English, you obtain 'week', which means a period of seven days, or 'weak', which is the opposite of 'strong'. That is the basis of language; from simple sounds, humans make ordered words. From words they can makes phrases, from phrases, sentences and so on. In the case of 'week' and 'weak', the context in which they are said is also very important and helps us to infer their meaning.

Language has been strongly implicated in the development of intelligence because it inherently teaches an individual to plan ahead. A random string of words will be meaningless, so some forethought is required before we speak. In everyday situations our responses are fairly routine but when we are trying to formulate new ideas and express ourselves as clearly as possible we sometimes have to slow down and think hard. Placing words in the correct order is known as 'syntax', although this term can equally apply to a set of ordered actions or thoughts. The saying 'be careful not to throw the baby out with the bath water' implies an incorrect syntax in a string of actions and is uttered as a caution to those who appear to be rushing into things without sufficient planning.

We learn language as children and it is universally observed that children understand language before they themselves can speak. In other words, a child can respond to a request before it can verbalise its own wishes. Scientists trying to teach animals to talk by pointing at symbols on boards are hoping to glimpse the development of the use of language. A child develops too rapidly to be very useful. The analogy has been drawn that in humans the old roads of language acquisition have been replaced by highways that allow a child to race to the finish line. So we study animals instead.

Chimpanzees and bonobos (sometimes called 'pygmy chimps') are, genetically, the animals closest to humans. It is an amazing fact that 98.5% of our DNA is exactly the same as a chimpanzee's DNA. Yet when was the last time you saw a chimp having a conversation at the bus stop? Also, remember that in Chapter 4 I stated that between 85% and 90% of our DNA is apparently useless. Therefore, the chances are that between 85% and 90% of the 1.5% difference between us and chimps occurs within the junk DNA. It is therefore possible that the differences between us and chimps are caused by just a few proteins. The problems facing geneticists at the moment are to precisely locate the differences and to determine how such a small difference in coding makes such a large difference in outcome.

It sounds to me as though intelligence could be another manifestation of emergence. Remember, I said in Chapter 4 that life emerges when a sufficient complexity of chemical interactions is achieved. So might it be with intelligence. When a brain reaches a certain level of complexity, it suddenly contains the machinery needed to become intelligent. Developing that machinery is the province of evolution, as I shall discuss in a moment.

From my point of view, the smaller the gap between humans and animals the better. If intelligence suddenly blossoms from a small genetic variation, then the more reasonable it is to believe that it might happen on other planets.

THE EVOLUTION OF INTELLIGENCE

To present the evolution of intelligence on Earth with a comprehensible analogy, many have likened the entire history of our planet to a single year. Ernst Mayr, an esteemed zoologist from Harvard University, represented the history of life's evolution something like this:

The formation of Earth	1 January
Prokaryotes	27 February
Eukaryotes	28 October
Chordates	17 November
Vertebrates	21 November
Mammals	12 December
Primates	26 December
Anthropoids	30 December at 1:00 am
Hominids	31 December at 10:00 am
Humans	31 December at 11:56:30 pm

Humans arrived on the scene only 3 ½ minutes before the end of the year. For most of the history of the Earth, there was nothing but simple prokaryotic cells. These are the cells without nuclei that I first discussed in Chapter 4. Mayr used this fact – and others that are based on the concept of contingency (that every evolutionary step is dependent on those before it) – to launch a damning attack on SETI. He simply could not believe that evolution would produce intelligence on another planet because he did not seem to believe that intelligence conferred any survival advantage to an organism. In fact, he described SETI as 'a deplorable waste of taxpayers' money.' Apart from that, of course, he could find nothing wrong with it.

Despite Mayr's formidable academic reputation, I am going to spend this section arguing against Mayr's point of view and I believe that there are now a number of neurophysiologists who would endorse my view that intelligence has indeed given us an evolutionary advantage. If it had not, how is it that we are at the top of the food chain?

I have already stated several times in this book that evolution is governed by the random flux of genetic mutations. When those mutations confer a positive survival strategy on the organism, the trait is passed on because the life-form lives long enough to find a mate and produce children. So, when considering the rise of intelligence, the question we need to ask is: 'Which survival problems does intelligence allow us to overcome?'

In the hand-waving definition of intelligence that I presented earlier, I said that intelligence is utilised in finding solutions to new problems. Another way of saying this is 'intelligence makes us versatile'. So in what way would versatility be an advantage? The answer is in a rapidly changing, perhaps even unstable, environment. This is why I stated in Chapter 6 that whereas a stable environment is good from the point of view of simply perpetuating life, a little instability stirs the pots and lets the versatile creatures flourish.

The Moon is a major factor in the Earth's environmental stability. Without its gravitational influence the axis of our Earth would oscillate in space by large amounts. At present, Earth's north pole is tilted towards the Sun by about 23°. Without the Moon, this would vary by significant amounts and cause some harsh climatic changes across our world.

John Barrow, of the University of Cambridge, and others, believe that these changes would be disastrous for life and that the Moon played a major factor in assisting life's development on Earth. Far be it for me to tread on such well-respected toes but I am going to stick my neck out and say that with reference to the subsequent development of intelligence, I suspect that the reverse may be true. Holding all other factors equal, the stabilising influence of the Moon may have delayed the emergence of versatile creatures.

The most basic requirement for any life-form is energy, as I discussed in Chapter 4. In plain English: all animals need to eat. If you eat only one single thing, bamboo shoots, for instance, and the bamboo crop fails, then you are in big trouble. I would like all giant pandas who are reading this book to take note and learn. Teenagers addicted to Big Macs could also learn a thing or two. How much easier it would be if they would simply expand their diet by finding some alternatives to eat.

Evolution is very good at breeding specialists who find an abundant food source and adapt to use it exclusively. This requires no intelligence; you simply live near your food source, so that when you wake up in the morning you start eating. That evening you hope your other half is in the mood for some love and then you go to sleep to prepare for another day of exactly the same tomorrow. When it comes to the perpetuation of your species, this arrangement is very efficient. If your environment changes and your food source dwindles, however, you are in big trouble unless you can adapt to eat something else. Chimpanzees are fairly versatile about what they eat: termites, fruit or even small monkeys are all on their menu. In a certain sense, giant pandas are narrow specialists, whereas chimpanzees are jacks-of-all-trades but masters of none.

Archaeologists can date the origin of our modern human brain to 2.5 million years ago. Half a million years after that, the fourfold expansion of the human brain was complete. It is believed that prior to this our ancestors possessed the same mental abilities as present day chimpanzees: just one typing sheet of cerebral cortex. The period of time during which this rapid development was taking place corresponds to a major period of environmental instability – an ice age.

There have been three great ice ages during Earth's history. One occurred around 2.5 billion years ago, another took place 800 million years ago and a third began 2 million years ago and might not yet be over, even though the ice caps are, at present, retreating. During an ice age, the average temperature of the planet fluctuates, with changes taking place over the course of decades. For long-lived animals this is far too short a period of time for evolution to adapt to the new regime, so versatile animals that can rapidly seek out new food sources or shelter will definitely have an evolutionary advantage. The more versatile the animal, the better they will be suited to a changing environment. Genetic mutations that promote versatility will be selected and, before you know it, hey presto! – intelligence. That's my theory, anyway.

I believe that environmental instability on a timescale comparable to the lifetime of the highest life-form present on the planet may promote its enhancement to a creature capable of intelligent thought. The first two great ice ages could not have had such an effect because during those periods (2.5 billion and 800 million years ago) life was still confined to the oceans.

Where does language fit into all this? Human beings are social animals, as are the chimps and bonobos. We all live in groups and most of us do our best to fit in. Knowing how others feel and what they want makes living in groups more difficult than roaming around on your own and therefore requires more intelligence. The more sophisticated the communication between individuals, the easier it is to assess the wishes of the group and plan group action. During hard times, communities have to pull together and share resources to survive. Those social creatures that are good at communicating would be better able to work together in order to survive. So again, adverse environmental conditions push us towards intelligence – this time by choosing syntax that can be used to communicate and plan ahead.

Another popular idea is that the ability to use tools drives the rise of intelligence. Tools would first and foremost be used to hunt because survival is always the first order of business. The ability to develop better tools would not be that advantageous if there were plenty of food to go around. If times are hard – as in an ice age – the ability to use tools to hunt effectively might be the difference between life and death. Perhaps the first tools were just stones to be thrown at animals. Even so, to throw a stone accurately requires the brain to perform an awful lot of complex planning. It must co-ordinate a complicated set of muscular movements in order to bring about the desired outcome.

Whichever way you approach this problem, you will be confronted by short-term environmental instability. As we have become more intelligent there has been a subtle but incredibly important, shift in our attitude towards the environment. We are no longer content to simply wait and see what is thrown at us. We try to understand and predict what will happen. There is nothing wrong in that except that sometimes we try to take preventive measures which result in making the situation worse. We have devised industry to make our lives easier, safer and more comfortable, yet the greenhouse gases that industry produces may be raising the temperature of the planet. Similarly, we are causing the destruction of our ozone layer. By trying to solve some environmental problems, we are simply creating others.

So can we, in fact, be too intelligent for our own good? There are those who certainly believe so and brand intelligence as nothing but a deadly pathogen. Intelligence, they assert, will inevitably lead not just to our deaths but to the catastrophic destruction of the Earth's ecosystem. Groups even exist that promote the peaceful extinction of humankind for the good of the planet. They are usually composed of individuals who have made a conscious decision not to have children and are urging others to do the same so that we can return the planet to the animal kingdom before it is too late.

DOOM SOON?

The idea that an intelligent species cannot last forever was acknowledged by Frank Drake in his list of factors that I summarised at the beginning of the last chapter. He believed that a certain fraction of intelligent species would not develop technology and even those that did might not live indefinitely. If a species became intelligent but lived in the oceans, that environment is hardly conducive to the discovery of fire and so technology itself might never develop. Even land-dwelling species might not manage to invent radio telescopes. After all, the ancient Egyptians established a high civilisation and were certainly intelligent but their priorities lay in the accumulation of wealth and beauty rather than in the pursuit of science.

There was a chilling reason why Frank Drake pondered the question of how long a civilisation would survive after developing radio technology. It was the fact that on Earth, radio technology was developed concurrently (more or less) with the atom bomb. For the first time during the four billion years that life had existed on Earth, one species now had the power to wipe itself out and obliterate many other species in the process. Worse than that, the Cold War between America and Russia made it appear as if that was precisely what was going to happen.

Our prospects certainly look better now that the Cold War has come to an end but we still face many uncertainties about our civilisation's future. Perhaps one of the greatest threats is posed by the collision of Earth with an asteroid or comet (see the illustration on p. 128).

It is widely thought that the dinosaurs were wiped out by the impact of an asteroid with the Yucatán peninsula in the Gulf of Mexico. The body responsible for the catastrophe is estimated to have been 10 km in diameter. It hit the Earth approximately 65 million years ago and wiped out most of the life on this planet.

The amazing truth is that it was not a unique event; there have been other mass extinctions throughout Earth's history. There have certainly been another four catastrophic events within the last 500 million years and some believe that there is compelling evidence for an extinction event roughly every 26 million years.

The good news is that, following an extinction, there is an explosion of life to fill all those now vacant environmental niches. The filling of niches by life-forms is termed 'biodiversity' and one epoch in Earth's history stands out as being head and shoulders above the rest in terms of biodiversification. It is known as the 'Cambrian explosion'. As far as life-forms go, Earth can be split into two periods. The Pre-Cambrian is the time between the formation of the Earth and 535 million years ago; in other words, about 4 billion years of history. For 3.5 billion of those, life was present on Earth but in primitive forms that lived in the ocean. About the most advanced living creature was a flatworm. The era has been termed the 'garden of Ediacara' because of a group of creatures, all long extinct, that lived in the oceans at the time. Then, suddenly, everything went crazy. Within 35 million years – the evolutionary blink of an eye – the number of species on this world literally exploded. According to the fossil record, this is the time at which the majority of the body shapes familiar to modern animals arose. This was the Cambrian explosion and it signified the beginning of our modern, animal era.

Theories abound as to what precipitated the Cambrian explosion. Some believe it was caused by the evolutionary development of genes that act like maps in developing organisms. The hox cluster of genes ensures that a developing organism knows to grow a head at the front and legs underneath.

Events Having No Likely Consequences	0	The likelihood of a collision is zero, or well below the chance that a random object of the same size will strike the Earth within the next few decades. This designation also applies to any small object that, in the event of a collision, is unlikely to reach the Earth's surface intact.
Events Meriting Careful Monitoring	1	The chance of collision is extremely unlikely; about the same as a random object of the same size striking the Earth within the next few decades.
	2	A somewhat close, but not unusual encounter. Collision is very unlikely.
Events Meriting Concern	3	A close encounter, with 1% or greater chance of a collision capable of causing localised destruction.
	4	A close encounter, with 1% or greater chance of a collision capable of causing regional devastation.
	5	A close encounter, with a significant threat of a collision capable of causing regional devastation.
Threatening Events	6	A close encounter, with a significant threat of a collision capable of causing a global catastrophe.
	7	A close encounter, with an extremely significant threat of a collision capable of causing a global catastrophe.
	8	A collision capable of causing localised destruction. Such events occur somewhere on Earth between once per 50 years and once per 1,000 years.
Certain Collisions	9	A collision capable of causing regional devastation. Such events occur between once per 1,000 years and once per 100,000 years.
	10	A collision capable of causing a global climatic catastrophe. Such events occur once per 100,000 years, or less often.

The Torino Asteroid Scale. As humankind devotes more time and resources to cataloguing the near-Earth asteroids, so a system for coding potentially lethal asteroids has been developed. At present we know of no asteroid that warrants more than a 1 on this scale (phew!) (Illustration courtesy Richard P. Binzel, MIT.)

Others believe that the Ediacara organisms lived in peaceful coexistence drawing sustenance from photosynthesising algae. Then a genetic mutation caused the first predator to appear on the scene. Suddenly there was a 'free-for-all', with many predatory species evolving because of the abundant free supply of food provided by the helpless Ediacarans.

So the creation of new species (defined as a group of animals that can interbreed) has taken place in fits and starts. The concept is known as 'punctuated equilibrium' and was proposed by biologist Stephen Jay Gould. Similar kinds of rapid development of species took place after each of the mass extinctions.

I believe strongly that the asteroid threat actually works in favour of intelligent species. Think of it as a back-up plan for any planet that is too environmentally stable to easily foster intelligence. Dumb animals simply adapt to this easy environment and fill every environmental niche. In that case, an asteroid collision every so often – to wipe them all out and let things start again – is a good idea. It wipes the slate clean and provides a lower form of life with the chance to evolve to fill all those now empty environmental niches – as did the mammals when the dinosaurs were eradicated.

The fact that every solar system in the Universe – not just in this Galaxy – will have formed from the collision of asteroid-like bodies, means that the threat of asteroid impacts will be a universal threat that must be faced by each and every inhabited planet. Only intelligent races will be able to develop science and thereby deduce the asteroid threat based on the fossil evidence for mass extinctions, the geological evidence of past impacts and the astronomical observation of asteroidal material. It is then up to them to do something about it. On Earth, there are many theories on the drawing board – such as the use of giant mirrors to focus sunlight on miscreant asteroids. This would melt part of the asteroid's surface, causing gases to be given off, which would act like a rocket to push the asteroid into a safe orbit.

In short, I think that the biggest threat to the continued existence of an intelligent civilisation may not be the threat of nuclear war but the threat from above: the indiscriminate rain of asteroids that have wreaked havoc on our world and will do so again in the future if we do not develop the infrastructure to combat stray asteroids.

HOW MANY INTELLIGENT SPECIES?

On Earth, there is a phenomenon called 'convergent evolution'. Its basis is that genetically distinct organisms evolve similar organs, body shapes or other properties to solve the same environmental problem. Perhaps the best example is the eye. Any creature that can see predators coming will stand a much better chance of surviving than something that is blind. Ernst Mayr (of hostile anti-SETI fame) showed that eyes have evolved at least 40 times on Earth, in completely different groups of animals. Mammals, insects and cephalopods, to name but three, have all developed eyes independently from one another. They all work in different ways but they are, nonetheless, all eyes.

Another example of convergent evolution is that of dolphins, sharks and the extinct icthyosaur. All evolved similar shapes and sizes to fit into the same environmental niche. So contingency can be avoided. Similar evolutionary solutions are possible, despite the different routes that organisms may take to reach them. Can intelligence also be like this? I believe that it can. Any planet with short-term environmental instability could evolve intelligent creatures. Intelligence is not an accidental by-product trait with no selective advantage; it has proved essential to our survival as a species. So, how many intelligent alien species can we expect?

In the previous chapter I presented Frank Drake's factors affecting the number of civilisations in the Galaxy. They can be represented as a mathematical equation that is basically a string of multiplication factors. When they are multiplied together with the communicating lifetime of a civilisation they produce an estimate of the number of civilisations that exist in the Galaxy. In mathematical symbols it looks like $N = R \times f_p \times n_e \times f_l \times f_i \times f_c \times L$. As a memory jogger, they are: R, the formation rate of suitable stars; f_p, the fraction of these stars with planets; n_e, the average number of potentially habitable planets; f_l, the fraction of those planets on which life forms; f_i, the fraction of those planets on which intelligence develops; f_c, the fraction of intelligent species that develop radio technology; and L, the average lifetime of a communicating civilisation.

As any SETI scientist knows, only one factor is truly important: the lifetime of the communicating civilisation. All the other factors multiply to about 1, depending upon the assumptions (sometimes blind guesses) which are made. So the lifetime is virtually equal to the number of communicating civilisations that will exist in the Galaxy at the same time as we do.

This lifetime does not necessarily need to be limited by the fact that the civilisation destroys itself in nuclear war or is rubbed out by an asteroid. Instead, it could simply mean that they develop more efficient means of communication and no longer use radio waves. Fibre optic cables, for example, are revolutionising the telecommunications industry on Earth. If their use signalled the decline of radio, Earth would begin to go quiet. So it is a radio-loud lifetime rather than a lifetime *per se*.

If the average radio-loud lifetime of a civilisation is only 100 years, SETI is almost certainly doomed to failure because the derived 100 civilisations will be spread randomly throughout the Galaxy with an average separation of about 10,000 light years. If a civilisation still leaks radio into space for 100,000 years, then SETI has a greater chance of being successful because calculations would lead you to expect the nearest civilisation to be at a distance of just 300 light years. When you consider that the Galaxy is 100,000 light years in diameter, 300 light years is virtually in our back garden.

At present it is impossible to tell which is the case. You are either a pessimist or an optimist. If you were a pessimist you would state that the facts are inconclusive and would read no more of this book. I, however, am an optimist, which means that I am going to plough on. In the next chapter I will talk about attempts to detect the little blighters.

Chapter Ten

Tuning-in to ET FM

I n the last chapter I described how language probably played a major role in the rise of human intelligence. Thinking about it, I am reminded of a cartoon I once saw, depicting two cavemen. One of them is saying to the others 'Now we have invented language, what shall we talk about?' Cavewomen, of course, had no such trouble and had already discussed their children's progress at school, decided on curtains for the cave and organised their first discussion group – but that is irrelevant. SETI scientists face a similar dilemma to our two hapless cavemen – with one difference. The caption should read: 'Now we have invented radio technology that can transmit across the cosmos, who shall we talk to?' Come to think of it, I don't recall ever seeing any SETI scientists wearing loin cloths either.

This chapter is all about using radio telescopes to search for radio signals emanating from alien civilisations. The previous chapters should have provided you with a feeling that, according to our modern thinking about the origin of life, intelligence and planetary habitability, aliens are out there somewhere. I think we have agonised enough about whether they might exist. It is now time to assume that they do and jolly well set about locating them.

THE START OF THE SEARCH

Speculation about the inhabitants of others worlds dates back to the Greek civilisation but the first suggestion that we could try to communicate with them took place in 1826. At that time it was assumed that beings lived on the Moon. The German mathematician Karl Friedrich Gauss suggested that the

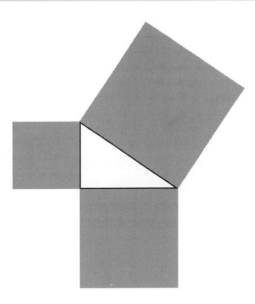

Karl Friedrich Gauss suggested that by cutting down enormous squares of forest in Siberia, Pythagoras' theorem could be represented in such a way that it would be visible to the Moon's inhabitants. This would let the Selenities know that Earth is inhabited by intelligent beings.

way to contact these Selenites (as they were known) was to draw a gigantic right-angled triangle to represent Pythagoras' theorem (see the illustration above). This could be achieved by cutting down enormous swaths of trees in Siberia. Another idea was to dig trenches in the Sahara desert and fill them full of kerosene. If the trenches were constructed in a circle and the kerosene lit, at night our planet would send an unmistakable geometrical signal into space. The Moon's inhabitants would see these symbolic representations as they studied Earth with their telescopes and realise that they were the product of intelligence. They would then make similar structures on the Moon.

A few decades later, in 1869, the Frenchman Charles Cros developed a communications code of flashing pulses and suggested that giant parabolic mirrors could be used to focus light towards Venus and Mars. Francis Galton – whom you met briefly in the previous chapter, when I described his setting-up of an intelligence-testing agency in South Kensington – also attempted to describe a mathematical language for communicating with extraterrestrials. He published this in 1896.

In the 1950s, however, the search for extraterrestrial intelligence took a turn which has dictated its progress for almost half a century. Cornell physicists Giuseppi Cocconi and Philip Morrison published a paper in a 1959 issue of the journal *Nature*, in which they described a scientific strategy designed to use the newly developed science of radio astronomy to listen out for alien signals. Note the subtle change of emphasis. Previous ideas had been to shout 'here we are!' but now these more conservative scientists sought simply to eavesdrop.

Central to Cocconi and Morrison's idea was the frequency at which to listen. Radio waves are perfect for transmitting signals over large interstellar distances because they pass through interstellar clouds without being dramatically affected. Light-rays, on the other hand, are blocked by such clouds. Looking at the radio region of the spectrum, the two authors came to the conclusion that the frequency to study was 1,420 Megahertz. This is the frequency at which hydrogen atoms naturally emit radio waves. It is an emission line of hydrogen. Specifically, it is known as the 'spin-flip transition line' because of the acrobatics that the electron around the hydrogen atom's nucleus has to perform in order to emit the radio wave. Since hydrogen is by far the most abundant element in the Universe, it was reasoned that 1,420 MHz – which corresponds to a wavelength of about 21 cm – would be a natural focus of attention for astronomers, both human and alien, throughout the Galaxy. One Megahertz is equal to one million Hertz, which means that a radio wave of 1,420 MHz is oscillating at 1,420 million times per second.

Any attempts to map the distribution of hydrogen in the Milky Way usually rely on observations at this frequency. So, would-be interstellar communicators might choose to transmit here because aliens might well be observing even if they are not consciously looking for intelligible signals.

At the same time as Cocconi and Morrison were busy developing the theory, Frank Drake was busy building a receiver to search for extraterrestrial signals. The frequency he had chosen, without any knowledge of the others' work, was also 1,420 MHz. Drake was not the first to use a radio telescope to search for radio signals from space but he was certainly the first to apply a strategic method to what he was doing. He constructed a receiver that would 'listen' to a narrow band of frequencies, about 50 Hz on either side of the 1,420 MHz hydrogen line. He then observed τ Ceti and ε Eridani, using the 85-foot diameter radio telescope at Green Bank, West Virginia. He observed for a total of 150 hours and heard nothing that could be interpreted as an extraterrestrial signal – but it was a start. SETI had begun.

Drake named the effort Project Ozma. Although it did not detect extraterrestrial intelligence, it *did* generate a lot of public interest. Even *Time* magazine ran a story on it – and that was where Bernard Oliver, Hewlett Packard's head of research, heard of the endeavour. Oliver became such a vocal advocate for the wide-scale funding of SETI, that when NASA commissioned a report into the feasibility of SETI, Oliver was asked to lead the investigation.

NASA BECOMES INTERESTED

Together with NASA's John Billingham (whose previous claim to fame was the invention of water-cooled underwear for astronauts), Oliver organised a multidisciplinary summer workshop at Stanford University which produced a landmark document entitled *A Design Study of a System for Detecting Extraterrestrial Intelligence*.

In one of the most influential parts of this report, Oliver wrote about where to look for signals in the electromagnetic spectrum. He noted that 300 MHz away from the hydrogen emission line is another emission line, produced by hydroxyl– a molecule consisting of a single atom of hydrogen and a single atom of oxygen (OH). By sheer coincidence, these two 'book-ends' exist in a part of the electromagnetic spectrum where background interference drops to a minimum. Oliver felt certain that since H and OH combine to make water, this naturally occurring 'waterhole' was the place to look for extraterrestrial signals. Just as waterholes on Earth are places of meeting and communication in the desert, he felt that the electromagnetic waterhole was the place for extraterrestrial civilisations to congregate and gossip. But where within the waterhole?

The waterhole is 300 MHz wide, whereas radio signals are transmitted in narrow bands of a few hundred Hertz at most. There are no known naturally occurring transmissions that have a bandwidth of less than 300 Hz. The report therefore recommended the development of computerised receivers that would be capable of splitting the waterhole into millions of narrow-frequency channels and analysing them simultaneously. This is the way that SETI has progressed.

As a brief interlude, I must mention that it is not just the Americans who believed in the search for extraterrestrial intelligence; the Russians were also captivated by the idea. They, however, opted for a different approach – one that assumed that 'super-civilisations' span the Galaxy. I have therefore decided to postpone my discussion of the beautifully romantic Russian ideas until the next chapter, which discusses the colonisation of the Galaxy.

Oliver's report contained an outline plan for something called Project Cyclops – an amazing project that had as its ultimate goal the construction of a vast array of 1,000 linked radio telescopes, each one of which had a diameter of 100 metres. Although the report called for the gradual development of such a vast system, the financial cost of the whole endeavour was on a par with the Apollo Moon-landings. Unfortunately, the Apollo programme had just been cancelled, due to withdrawal of funding.

Project Cyclops was far too big to be feasible but NASA's interest in SETI would not go away. Some members of the US Congress, however, did not share

NASA's enthusiasm for the endeavour. It seemed that no matter how small the proposals were, there was a constant battle for government funds.

OUT OF THE ASHES

NASA doggedly continued to fund a SETI development programme at subsistence level, with workers being located at NASA's Ames Research Center at Mountain View, California, and the Jet Propulsion Laboratory at Pasadena, California. Working together, they proposed a dual approach. The people at Ames would target their search, selecting 1,000 stars that are similar to the Sun, for close scrutiny. The JPL group would conduct a more general, though less sensitive, survey of the entire sky.

In 1988, NASA formally approved these two projects and began gearing-up to do some serious science. The project was led by stalwarts Oliver and Billingham and by Jill Tarter, an astronomer who had developed a passionate belief in the existence of extraterrestrials. On 12 October 1992, NASA began what it had by now termed the High Resolution Microwave Survey (HRMS). It had previously been called the Microwave Observing Project but presumably someone decided that such a high-tech endeavour was probably not best served by saying that a team of scientists were sweeping the sky with a MOP.

Almost one year later, Congress finally, irrevocably and quite specifically, withdrew funding from SETI. The man responsible for this turn of events was Nevada Senator Richard Bryan. Despite the fact that HRMS was less than 0.1% of NASA's annual budget – amounting to about a nickel per American taxpayer per year – the Senator successfully argued that it cost too much.

Out of this disastrous turn of events rose the SETI Institute. It resurrected Ames' targeted search, expanded it and called it Project Phoenix after the mythical bird that rose from its own ashes. Project Phoenix is a continuing search programme that is now funded entirely by private donations. It has searched for extraterrestrial signals at radio observatories around the world, including Arecibo in Puerto Rico, Parkes in Australia, Jodrell Bank in England and Green Bank in America. It is amazing to note that Project Phoenix is estimated to be 100 trillion times more sensitive than Frank Drake's Project Ozma.

A key aspect of Project Phoenix is that the data are processed in real time. In other words, you need a computer fast enough to look at the data, there and then, so that you can be alerted immediately if anything warrants further investigation. That way, if a promising signal is detected, other observatories can be contacted immediately so that they can ascertain whether they, too, can pick up the signal.

This is the rationale behind the simultaneous observing programme run by the SETI Institute at Arecibo and Jodrell Bank. If a signal is detected by Arecibo, it can be immediately followed up by Jodrell Bank. When Project Phoenix observed for a total of 24 weeks over the 19 months between September 1996 and April 1998, using the 140-foot telescope at Green Bank, it utilised a second telescope in Woodbury, Georgia to verify candidate signals.

OTHER SEARCHES

NASA and the SETI Institute are by no means the only institutions to take up the challenge of SETI. The University of California at Berkeley, and Harvard University, are instrumental in searching. Most necessarily, a private organisation called the Planetary Society is instrumental in funding these searches.

The Planetary Society was co-founded by the late Carl Sagan as far back as 1980. It sought to promote the exploration of the Solar System and the search for extraterrestrial life and intelligence. It is a private, non-profit making organisation that ploughs its money into research and increasing the public awareness of space. It is the largest organisation of space enthusiasts on this planet, with more than 100,000 members in 140 countries. By 1999, the Planetary Society had given over $1 million to SETI searches.

Harvard physicist Paul Horowitz became interested in SETI late in the 1970s. By 1981, he had his own plans to build a portable receiver that could be transported easily to whichever radio telescope would have it. With modest funding from NASA and the newly-formed Planetary Society, Horowitz built this receiver and called it Suitcase SETI. His receiver analysed the radio waves in very narrow wavebands so that he could eliminate as much noise as possible. This meant however, that to cover a reasonable range of frequencies, his receiver needed to be able to simultaneously analyse a very large number of these ultranarrowband channels.

Suitcase SETI looked at 64,000 channels and then in the mid-1980s the film producer Steven Spielberg made it possible for Horowitz to construct an 8-million channel receiver by donating $100,000. This led to the construction of META – the Megachannel Extraterrestrial Assay. A version of META, known as META II, is currently running in Argentina under the guidance of Guillermo A. Lemarchand of the University of Buenos Aires. Horowitz's greatest achievement to date has been the construction of BETA, the Billion-channel Extraterrestrial Assay. This amazing device (which really has only a quarter of a billion channels) can scan the entire waterhole in ultranarrowbands. It is currently on-line and is searching for that elusive whisper.

Another project is SERENDIP – the Search for Extraterrestrial Radio Emissions from Nearby Developed Intelligent Populations. The name plays on the word 'serendipity' – an unexpected and accidental discovery. This project is run by C. Stuart Bowyer and Dan Werthimer, from the University of California at Berkeley. They constructed a device that will listen at 1,420 MHz in whatever part of the sky the Arecibo Radio Telescope happens to be facing while it is working on other projects.

SERENDIP has been running for 18 years and in June 1997 was upgraded to enable it to analyse signals from 140 million frequency channels in just 1.7 seconds. Project SERENDIP has created so much data that it has been impossible to fully analyse them all in real time. To tap into the power of the home computer revolution, computer scientists conceived a plan to write a screen-saver program. Screen-savers are designed to protect monitors by providing ever-changing patterns rather than an unchanging image being frozen on the screen when the computer is not being used. The SETI screen-saver would download a chunk of information from the Internet, analyse it thoroughly, flag any interesting signals and then return the data to the main SETI computer.

The project is known as SETI@home and is available for immediate download from the Internet. In its first week of operation, just under 300,000 computers were loaded with the software, which resulted in 9.5 million hours (1,100 years) of computer processing power being brought to bear on the challenge of finding ET. At the time of writing there were more than one and a half million people running SETI@home in 224 countries. Together, they had contributed 125,000 years of computer processing time but as yet had detected nothing. In fact, most of this book has been written on my trusty Mac, which has SETI@home installed as the screen-saver. Who knows, it may be my machine that spots the first detected signal.

Astronomy still relies on the vast contribution made by amateurs to assist the professionals. Without doubt, this is one of the nicest parts about the discipline. As the example above illustrates, SETI is no exception. How fitting that the search for extraterrestrial intelligence should rely on a global effort from ordinary individuals around the world, rather than just a few academics in lofty ivory towers. To underscore this, another non-profit-making organisation – the SETI League, under the leadership of Paul Shuch – is trying to co-ordinate a global effort that uses 5,000 small dish antennae in 40 countries to provide constant coverage of the whole sky.

The year 1999 saw the University of California at Berkeley team up with the SETI Institute in a joint venture to construct a array of 500–1,000 small radio telescopes that can simulate one big dish. To underline their commitment to this project, Berkeley created a chair in SETI and awarded it to William Welch.

He will direct the construction of the $25 million radio telescope, which will be known as the 1-hectare Telescope (1hT) and will eventually have a collecting area of 10,000 square metres. This revolutionary instrument is designed to prove that many small telescopes can do the work of one giant instrument for a fraction of the cost.

The 1hT will be substantially devoted to SETI, although other radio astronomy will be performed with it. If all goes to plan, the instrument could be completed by 2004 and, instead of Project Phoenix's current rate of a few hundred stars a year, 1hT will enable Welch's team to scan a few thousand stars every year.

FALSE ALARMS AND ANNOUNCING THE DISCOVERY

Although none of the SETI searches have detected an extraterrestrial message, all have produced candidate signals. SERENDIP reports about one candidate signal every few days, whilst the META project turned up dozens of interesting candidates. In 1977, a SETI search run by Ohio State University detected a signal that almost went off the power scale. Project scientist Jerry Ehman wrote 'Wow!' next to it when he saw the print-outs. It has since become immortalised in SETI mythology as 'the WOW signal'.

If a signal is to be shown without doubt as being extraterrestrial in origin, it must be reobserved. It is similar to catching a glimpse of a shadowy figure in a dark room. Our first instinct is to look again to make sure we really did see something. If, on the second look, the SETI scientists detect the signal again, then they must begin the arduous task of establishing whether it is extraterrestrial in origin. I cannot stress enough that at the time of writing this book, no candidate signals have been redetected and shown to be from extraterrestrial intelligence.

Periodically there are news stories about the possible detection of alien signals. For example, on 22 October 1998, BBC News Online ran a story about an amateur radio astronomer who claimed to have picked up a message from the star EQ Pegasi, just 22 light years away. Project Phoenix had observed the star a month before and detected signals. Before making any announcement themselves, they had run their routine checks and determined that they were terrestrial interference. The story swiftly died.

Some individuals and groups have used stories such as this to argue that the discoveries are actually covered up. A few even believe that the SETI scientists are even now in direct communication with aliens. The reality is that most SETI projects struggle to survive on private donations. If they really did have proof of the existence of extraterrestrials, SETI would become the biggest scientific discipline in the world. Everyone would want a piece of that pie and, before you knew it, Project Cyclops would be back on the table. The same argument goes for people who believe that NASA has covered up the evidence of a martian civilisation. If there were such evidence, I am confident that production would begin tomorrow for a human mission to Mars.

The truth – unpalatable as it may be to some – is that there is as yet no evidence for the existence of extraterrestrial life. But that is no reason to give up searching because, to quote a SETI cliché, 'absence of evidence is not evidence of absence'. Paul Horowitz summed this up beautifully at the inauguration of the BETA search when he promised to jump from the top of the radio telescope if someone in the audience could prove that extraterrestrials do not exist.

On that hallowed day when something is found, everyone will know about it. There is a set of applicable principles which has been endorsed by six international space organisations. These principles govern the verification and the announcement of the detection of an extraterrestrial signal. They state that the discovery of a signal that is potentially extraterrestrial in origin must be promptly communicated to other observers and research institutions that are signatories of the same set of principles. This will allow the signal to be confirmed. If the signal cannot be confirmed as coming from extraterrestrials, then all the original discover is allowed to do is claim detection of an unknown phenomenon.

If the signal is confirmed as coming from an extraterrestrial intelligence, then the discovery must be circulated through the International Astronomical Union – as are the discoveries of new celestial events or objects, such as comets or supernovae. The signal data must be provided to a number of top telecommunications and astronomical research centres throughout the world. Also, the Secretary General of the United Nations and the relevant national authorities must be informed.

The original discoverer also has the privilege of announcing the confirmed detection of an extraterrestrial signal as promptly as possible to the general public and scientific community at large. The International Telecommunication Union may request that Earthly transmissions be minimised at the frequency of detection in order to facilitate the signal's subsequent reception and study.

However, any decision about sending a reply is strictly forbidden until an international committee of astronomers has been convened to review the entire situation and decide whether a response is appropriate.

In short, it would be next to impossible for a government to cover up any such detection. As well as the declaration of these principles, simple celestial mechanics would virtually guarantee worldwide reception. This is because the Earth rotates once every 24 hours. If the signal is to be tracked it will need a string of radio telescopes placed around the world so that as one loses sight of the radio's source another will be able to pick it up.

One of the fascinating questions to ask about the whole issue of life on other worlds is: what would humankind make of such a detection? The presence of microbes on Mars might be easy to assimilate but what of the knowledge that an intelligent alien civilisation exists somewhere else? Those reading this book are probably pro-life and would react with enthusiasm to such knowledge. No one can doubt, however, that there are those who would be horrified and panic-stricken by the news.

Looking back into history, there have been a number of defining moments when humankind has been forced to change its view of the Universe by a large factor. Such radical changes are often referred to as 'paradigm shifts'. One such paradigm shift was caused by Copernicus and Galileo when they advocated that the Earth was not the centre of the Solar System. Another was Charles Darwin's theories on the origin of species. Each was heavily opposed by a vocal minority but nevertheless was gradually accepted. Perhaps the same will happen with the detection of extraterrestrial life.

Doug Vakoch, of the SETI Institute, is a social scientist who has studied the possible effects of a successful detection by interviewing people about their views on SETI. He discovered that, as a broad generalisation, the people most opposed to the idea of aliens are the religious fundamentalists. It is therefore these people who are likely to be the most outspoken detractors if a signal is found.

A NASA report on the issue of society's response to an alien transmission made the point that terrorist activities could not be ruled out. Perhaps these would be directed at the establishments responsible for the detection – or even against the individual scientists themselves.

Although it is impossible to predict behaviour, it seems likely that most people will simply be curious to know more. However, the brutal truth is that the signal may not be decodable. Part of it may be missing, or it will be so alien that we cannot get our heads around it. We will simply be able to recognise that it came from a transmitter, rather than from a natural source. In time, it may be cracked – but who knows?

Perhaps this is the best way to start. A simple confirmation of life elsewhere is unlikely to start a global panic, whereas a signal that says 'We're on our way; put the kettle on' would almost certainly have a very different effect on humankind.

OPTICAL SEARCHES

It is not just radio searches that are continuing. In 1999, astronomers finally realised that optical signalling methods are possible. The chief advocate of this approach to SETI was Charles Townes, the inventor of the laser. When he developed lasers in the 1950s, he realised that they would be very powerful signalling devices and expounded his ideas. Although lasers were ultimately taken very seriously – to the extent that they are now in every supermarket to read bar-codes and every CD and CD-ROM player – Townes' SETI ideas were not. In 1964 he won a Nobel Prize, along with two Russians who had independently developed the principle of the laser. As a final vindication of his foresight, at the age of 83 Townes has now seen the initiation of searches for optical pulses from nearby stars.

Whilst it is true that communication over vast distances would be severely limited by intervening clouds of dust and gas, it is simply common sense that if astronomers can see a star in a telescope, then in principle they could also see an optical signalling device. The signal would most probably be a pulsing beam of light, rather similar to the way messages in Morse code were transmitted between sailing ships using flashing lights.

Three searches for optical signals were initiated in 1999, thanks to generous donations of money from The Planetary Society and the SETI Institute. Two of the searches are being co-ordinated at the University of California at Berkeley. The third is being conducted at Harvard University and the Smithsonian Radio Observatory in Massachusetts.

Of the two Berkeley searches, one is using the backlog of data generated by Geoffrey Marcy in his trail-blazing effort to detect planets around other stars. The other is using a telescope to search for pulses that may last for just a billionth of a second. The receiver necessary to do this has been duplicated by the Harvard team and is currently 'piggy-backing' on an experiment that is measuring the velocity of 2,500 nearby Sun-like stars.

ACTIVE SETI

Despite the fact that Project Phoenix is so sensitive, it is still estimated to be incapable of picking up stray radio emission, such as television or radio broadcasts, from technological civilisations. That was what Project Cyclops was designed to do. Instead, all SETI searches so far rely on extraterrestrials purposely trying to draw attention to themselves. In its most extreme case, they believe in the existence of the Galactic Club (which I discussed in Chapter 1). This is envisaged to be a communicating network of alien worlds that canvas for new member civilisations by broadcasting powerful radio signals into space. Think of it as a kind of interstellar junk mail – but instead of the local gym trying persuade you to join, it would be the member races of the local galactic civilisation.

As I stated in Chapter 1, I do not personally like this assumption. The less we try to second-guess what ET is up to, the more comfortable I am. What I would like to see is every radio telescope fitted with a receiver, rather like SERENDIP, that automatically takes the radio data being collected and checks for the evidence of signals. It would not matter what frequency or object the practising astronomer were looking at. An automatic computer check would sneak a peek – just in case. The inescapable fact is that more than 99% of the radio observations undertaken to study the Universe are not SETI searches. So how wonderful it would be if a little bolt-on device could make use of all these observations, rather like the way in which two of the optical SETI searches make use of data being taken for other, more conventional, science programmes.

By this, do not for one minute believe that I am consigning the current SETI searches to being a waste of time. Far from it. It is the expertise of the people currently involved in them that will make building such 'fly-on-the-wall' receivers possible. It will be their specialised knowledge that will determine what should be done when something is detected. SETI scientists are the astronomers that set the observational benchmark for all other astronomers. As one astronomer said to me after watching SETI people in action at Jodrell Bank: 'If we applied the same levels of data validation to our observations as they do to theirs, we would NEVER publish any results.' Imagine if your own data had the potential to create a watershed in human history. You would be very, very careful indeed about drawing the wrong conclusion from it.

Guillermo A. Lemarchand believes that the first alien signal will not be detected by a SETI astronomer but by a radio astronomer of the common or garden variety. They will be studying something completely different and buried in their data will be a signal that they cannot fathom. After eliminating all natural explanations they will be forced to take it to a SETI scientist: *et voila*! – first contact will be made. Lemarchand advocates that we should just wait and see what happens.

So we listen ... and just hope that it is the aliens who are transmitting. What cosmic irony it would be if the Universe is chock full of intelligent civilisations, all scared to make deliberate signals but all desperately listening for the sounds of the others. It conjures up images of two eavesdroppers, one on each side of a closed door, desperately listening but totally unaware of each other's presence.

Other astronomers do not wish to simply sit around and wait. They want to

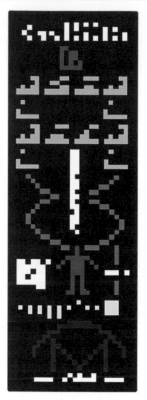

Frank Drake's interstellar message was sent from Arecibo in 1974, in a communication that lasted for just three minutes. If aliens apply the same redetection criterion to verifying potential signals that SETI scientists on Earth do, they will be forced to discount the message because the transmission was never repeated. (Illustration courtesy F. Drake (UCSC) et al., Arecibo Observatory (Cornell, NAIC).)

start shouting again. This brings us back to the concept of the test for echo that I began with in Chapter 1. SETI pioneer Frank Drake did not spend all his time just listening. In 1974 he used the Arecibo Radio Telescope to transmit a message towards the star cluster M13, about 25,000 light years away. It was a simple dot pattern that made up a diagram of the Solar System, a human being, DNA and a radio telescope (see the illustration above). Critics said that it was largely symbolic and stood no real chance of being detected but, conversely, it sent some people into a xenophobic rage as they wondered whether laws should be passed to ban such practices.

England's then Astronomer Royal, Martin Ryle, was concerned enough to write to Frank Drake and question the wisdom of making ourselves known. Drake's reply was to point out that Earth had been leaking radio waves into space for years, so we were already unavoidably advertising our presence.

Central to the concerns of those who have objections to transmitting our presence is the fear that invading armies of aliens will arrive to conquer and enslave us. Other, more optimistic, folk believe that upon contact we will be given vast quantities of knowledge by benevolent aliens. I suspect that this might, in fact, be as disastrous as the invasion scenario. Giving anyone too much knowledge before they are mature enough to deal with it is always a bad move.

Perhaps *Star Trek*'s creator, Gene Roddenberry, was right when he created the Prime Directive for his internationally famous space exploration television show. According to the Prime Directive, we should not contact aliens until they themselves are on the verge of interstellar space-flight. Roddenberry once said that he felt it was good that solar systems are so far apart that they require enormous technological sophistication even to contemplate a journey between them. He went on to explain that the reason it was good is because it means we would make all our errors in our own Solar System and not annoy anyone outside. By the time we are ready for space-flight, perhaps as a species we will have matured enough to behave ourselves.

The most recent attempt to contact aliens has been masterminded by French Canadians Yvan Dutil and Stephane Dumás. When they began thinking about constructing a message they discovered that a universal language had been proposed by Dutch mathematician Hans Freudenthal in 1960. Dutil went to the science library at the Université Laval to borrow the book, *Lincos: Design of a Language for a Cosmic Intercourse*. To his surprise, he was the first person to borrow the book for 36 years. He claims that it soon became obvious why the book had gathered dust for so long – 'It bores you to death!'

Dutil, however, persevered with the tome, learning much about structuring messages so as to be as clear as possible. Ultimately though, he and Dumás decided that they would send pictures rather than a message totally made up of words. Across a total of 23 pages, the pictorial message depicts scientific concepts such as Pythagoras' theorem, the structure of hydrogen and helium, the schematics of DNA, planet Earth, the Solar System and a great many other things (see the illustration on p. 146) The four-hour message was transmitted at a wavelength of 6 cm on 24 May 1999, from the Evpatoria dish in the former Soviet Union.

The transmission was financed by an American company called Encounter 2001, which offers to beam personal messages into space for $14.95. Public relations companies also see this as the ultimate publicity stunt. How better to draw attention to yourself on Earth than to announce yourself to the stars. It seems like a silly gag that does no one any harm. But what if aliens are listening? What if they believe they are hearing the Earth's representative and that our word is our unbreakable bond?

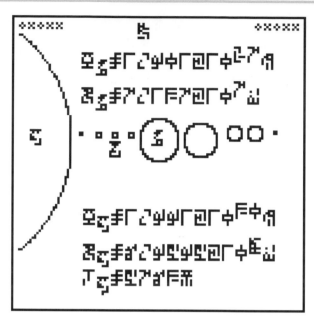

A page from Yvan Dutil and Stephane Dumás' message. This page depicts the Solar System in a schematic form. (Illustration courtesy Yvan Dutil and Stephane Dumás.)

Transmission of messages is an area of space law that needs considerable tightening and, more importantly, subsequent enforcement. No matter how humankind chooses to tackle the issue of purposeful transmission into space, as Frank Drake wrote many years ago, whether we like it or not, Earth is a radio-loud civilisation. Television and radio transmissions leak into space every second of every day. All we can hope for is that the film *Independence Day* really was just fiction ...

Having drifted a little into science fiction country, let us continue this line of thought. In Chapter 1 I gave my reasons for believing that, in time, the stuff of science fiction – such as so-called superscience and hyperdrives – would become possible. This forces me to consider that some aliens somewhere have already progressed sufficiently to make these discoveries. With this thought in mind I want to spend the next chapter thinking about whether the Galaxy already has an established Galactic civilisation. Think of it! Spacefaring species could be travelling between the stars at this very moment, trading and living with each other, rather as we flit from country to country on Earth. They are probably even whistling the theme tune to *Star Trek*.

So let us move on and think about the existence of a Galactic infrastructure, peopled by super-civilisations that are as far in advance of us as we are from chimpanzees.

Chapter Eleven

The odds of receiving

a visit

Some individuals wonder why astronomers bother to search for radio signals. They consider that it would be simpler to send off a spaceprobe in search of extraterrestrials. Oh, that it were so easy!

All of the spaceprobes we hear so much about are missions that take place within our own Solar System. The star nearest to our Sun is the triple star system α Centauri – otherwise known as Rigel Kentaurus. It lies at a distance of 4.3 light years, which means that the light it emits takes 4.3 years to reach us. Light is the fastest thing in the Universe. Everything else travels more slowly.

The fastest rockets travel at about 10 miles (16 kilometres) per second. It takes me just over half an hour to drive from my home to the University of Hertfordshire but in a rocket I could do it in less than two seconds – so a rocket seems pretty fast. Indeed, rockets and the spaceprobes that they carry are the fastest man-made objects. In comparison to the speed of light, however, this is nothing. Light travels at a speed of 300,000 kilometres per second – which makes 16 kilometres per second seem a little feeble.

TO BOLDLY GO ...

Think about α Centauri. A conventional spaceprobe would take about 60,000 years to reach it. We have no experience of building an electronic device that could work for that length of time on Earth, let alone in space. Spaceprobes are notoriously temperamental things because of their complexity and because they are designed to work in the harsh radiation-filled environment of space.

Within a radius of about 100 light years from us, there are approximately 1,000 stars like the Sun. Investigating all of them with spaceprobes would require an incredible investment of time, effort and, most importantly, money. Scientists would begin working on projects from which their 2,700th generation of descendants would reap the benefits. However, what if the time taken for a spaceprobe to reach another star system could be reduced to the working lifetime of an average human?

Now we are really cooking! Imagine being a graduate student in your early 20s, working on the final preparation of the mission. It is launched as you are awarded your PhD at 25. Forty years later – when you are a distinguished professor at a prestigious university – the spaceprobe arrives at α Centauri and begins to relay back images and data. It is the crowning glory of your career and you retire at 70, having just completed your best-selling memoirs.

If the target system is α Centauri, then the speed our spaceprobe needs to achieve is at least 10% of the speed of light – about 2,000 times faster than the fastest rocket today. A 1998 NASA study, co-ordinated by Henry M. Harris of the Jet Propulsion Laboratory, has indicated that such speeds are achievable and he has identified three methods that could work.

The first method is nuclear fusion. This is the process that takes place inside the hearts of stars (as I explained in Chapter 3). In outline, two particles are collided with such force that they fuse together and release energy. The particles that fuse and the particle they fuse into are electrically charged. This means that strong magnetic fields could direct the particles out of the back of the starship – just like a rocket exhaust – and thereby push it forward.

The easiest fusion reaction to initiate relies on two isotopes of hydrogen: deuterium and tritium. Each is a normal nucleus of hydrogen (which is just a single proton) but deuterium contains an additional neutron, whilst tritium contains two additional neutrons. These neutrons actually cause a problem because, as their name implies, they are electrically neutral and so cannot be controlled by magnetic fields.

Although a nuclear fusion powered starship would get you to α Centauri, you would have been fried by the neutrons before you left our Solar System. Hmmm. Slight design problem, methinks. Either some pretty hefty shielding, or a different fuel that does not generate the neutrons, would be essential. It is possible to substitute the tritium for a light isotope of helium that is deficient by one neutron. In the fusion reaction, the deuterium gives it a neutron and a proton is ejected. The proton, like the resultant helium nucleus, is then controlled using the magnetic field. Light helium – helium 3 – is incredibly rare on Earth but the surface of the Moon, acting like a sponge, has been soaking them up for billions of years. In order to get at them, there would have to be a substantial mining effort in place on the lunar surface.

This may not be as far-fetched as it sounds. It is almost inconceivable that interstellar ships will launch from the surface of the Earth. Instead, they will be built in space and their engines will be turned on to first nudge them out of Earth orbit and then to send them on their way to the stars. So substantial near-Earth infrastructure will have to be in orbit first. This is why the International Space Station is such a crucial step. Improving our construction abilities in space and building on the work of the Russian Mir space station programme to maintain a permanent human presence in space, is of vital importance if we are ever to reach for the stars.

With a fusion rocket it may not be essential to carry all the fuel on-board. A gigantic magnetic field could be utilised to funnel fusionable material out of the incredibly tenuous gases that exist in space, into the spacecraft. This concept was first proposed by Robert Bussard way back in 1960 and has become known as the 'Bussard ramjet'.

To place the power of fusion into some kind of context, we can compare the amount of energy a kilogram of fusion fuel would release, with that of a kilogram of chemical rocket fuel. It is a staggering 10 million times greater. If you think that this is impressive, the 1998 NASA report also identified the second possible propulsion method as being the annihilation of matter with antimatter. Remember Chapter 2, in which I described antimatter as being nature's way of balancing the books just about every time an energetic reaction creates particles. The flip side of this is that when a particle meets its antiparticle counterpart, there is one almighty bang as they instantly annihilate each other.

A kilogram of matter and antimatter would liberate about 200 times more energy than the fusion fuel, making it 2 billion times more efficient than chemical rockets. This same amount of matter and antimatter would produce enough energy to supply the entire Earth's requirement for 26 minutes.

Creating and handling antimatter is (currently) incredibly difficult. Only in particle accelerators can antimatter be produced. Every year, the world's production of antimatter in these facilities amounts to a few hundreds of millionths of a gram. This is not much. Considering that it annihilates on contact with normal matter, you will need something much more sophisticated than a jam-jar in which to catch it.

The good news is that if you create antiprotons, they are negatively charged and so can be controlled with magnetic fields. There are several research groups around the world who are presently designing magnetic 'bottles' for antimatter.

The third candidate for powering a starship is a very elegant concept. It is known as a 'solar sail'. Basically, it is a gigantic, thin mirror that is unfurled by a spacecraft. The light from the Sun reflects off the mirror and produces a recoil which pushes the solar sail along. The beauty of the idea is that the ship needs no fuel and no bulky engines; nor does it need dangerous stuff such as antimatter.

However, the further from the Sun the solar sail travels, the more feeble is the push from the light. To solve this problem, you simply(!) construct laser platforms in space. These take energy from giant solar panels and convert it into laser light. Aim this at your solar sail and off it goes. It is the optical equivalent of blowing in the sails of a model ship.

It was the concept of a solar sail that won the prize in the 1998 NASA report for the most feasible concept to develop in the near-term. To this end, on 8 March 1999 Sarah Gavit was appointed the Associate Manager for the Interstellar and Solar Sail Program. Current plans call for a test solar sail to be launched as soon as possible.

As technologically daunting as all this talk of starships might sound, I believe that the biggest problem standing in the way of interstellar flight is the same as the obstacle outlined by Eugene Mallove and Gregory Matloff in 1989, in their seminal book, *The Starflight Handbook*: '... Interstellar travel may, indeed, be virtually impossible, if the self-defeating assumptions are put in the way.' In other words, if we simply believe that we can do it, then the finest human minds will find a way to make it happen. I think there is a growing realisation at NASA that this is indeed the case. I predict that the first century of the third millennium will be recorded in history as being the first time humankind sent a spaceprobe to the stars.

ARE ALIENS EXPLORING THE GALAXY NOW?

If such things as interstellar travel are possible, then presumably ETs are up there now, boldly going where no fuzzy little creature from planet Xon has gone before. This is the assumption that many Russian scientists made when they became interested in SETI during the middle decades of the twentieth century.

There is a remarkable little book tucked away in the University of Hertfordshire's library. It is a slim volume of 100 pages, in a nondescript brown fabric cover. Despite this, it is a little piece of SETI history. It is the Proceedings of the First All-Union Conference on Extraterrestrial Civilisations and Interstellar Communication, held at the Byurakan Astrophysical Observatory of the Armenian Academy of Science in May 1964. In the book, the conference delegates published papers that talked of 'super-civilisations'. Nicolai Kardashev, of Moscow State University, classified three types of civilisation in his contribution. The criterion he used to differentiate the civilisations was the rate at which they consumed energy. He stated that the first level of a civilisation is that in which Earth presently finds itself. It is utilising the power output available on its own planet, including the Sun's radiation that is intercepted by Earth.

The second type of civilisation is one that is using almost the entire power output of its primary star. In order to do this, space stations and orbital colonies would have to literally surround the star, intercepting as much of the precious radiation as possible. The concept of a spherical swarm of artificial constructions around a star is known as a 'Dyson sphere', after British physicist Freeman Dyson, who suggested the possibility in 1959. Dyson had in turn taken his inspiration from Olaf Stapledon who, in his book *The Star Maker*, published in 1937, suggested the concept as a way to live around a planetless star. It has been somewhat misconstrued in science fiction and is often depicted as a solid construction that completely encloses the central star.

Although both approaches would effectively block out the light from the central star, the waste heat produced by the industrial and other processes might be expected to radiate away from the sphere as infrared radiation. Dyson estimated that this might have a peak wavelength of about 10 μm. So far, no infrared observations have shown anything that cannot be explained as a naturally occurring celestial object – but who knows what may be discovered tomorrow?

Kardashev's third type of civilisation was one that had harnessed the entire power output of a galaxy. Effectively, they would need to construct Dyson spheres around almost every star in the galaxy. Well, it is quite obvious from simply looking into the sky that the twinkling stars prove there are no Kardashev Type III civilisations currently rampaging through our part of the Milky Way.

Early in the third millennium, if present consumption rates grow as expected, Earth's energy needs will outweigh the resources available from our planet. As a result, we will be forced into space if our civilisation is to keep growing. Kardashev believed that as a civilisation grew, so the energy of its transmitters would grow because of the necessity of communicating over increasingly large distances.

I am still not convinced that radio waves are the beginning and end of communication. As a speculative example I suggest the following, which my colleague Jim Collett and I devised over lunch (the university must have slipped something into the coffee that day). Note that every time a scientist uses the word 'speculative', he/she has no hard mathematics to underpin the idea – just a hunch; so I am not claiming that this idea is workable but every idea has to start somewhere.

Jim and I wondered whether an instantaneous communications device is, in principle, possible. We wondered this because there are several experiments that seem to imply information is being instantaneously communicated between particles. It relies on an effect called 'quantum entanglement', with which physicists are just getting to grips. Pairs of entangled particles are inextricably linked, no matter how far away they are from each other. One of the leading physicists in this field, Jeff Kimble, of the California Institute of Technology, expressed it thus: 'Entanglement means if you tickle one, the other one laughs.' The most important point is that this information is communicated instantaneously. What a thought! Real-time communication across galactic distances would be possible because there would be no time delay as there is when radio waves travel through space at the speed of light.

Entanglement is a crucial aspect of quantum information theory and researchers in the field are working hard to perfect their ideas so that quantum computers can be built which would make the most powerful computers today look like Clive Sinclair's pioneering home computer, the ZX-81.

Our ideas on entangled communication are far from being rigorous science but as a concept I think it is at least worth pursuing. I have also used it as a central theme to the science fiction novel I am writing at the moment. During the course of research for the book you are reading, I discovered that John Walsh, of the SETI Research and Community Development Institute in Brisbane, has had similar ideas about instantaneous communication. Sougato Bose, of Imperial College London, won the Wellcome Trust/*New Scientist* 1999 Millennial Essay Competition for postgraduate students with a discussion of entanglement as a means of communicating in total secrecy. I suspect that others have also thought along such lines – even if only quietly.

In the final section of this chapter, I shall return to these ideas when I consider 'The Great Silence'.

SENDING ROBOTS AHEAD

Judging by the pattern of Solar System exploration, robotic spaceprobes will certainly be sent before crewed starships. In fact, all of NASA's interstellar aspirations may be geared towards capitalising on the Darwin and the Terrestrial Planet Finder missions, discussed in Chapter 8, to search for metabolic gases in the atmospheres of nearby Earth-sized planets. As soon as an interesting system is found, NASA will accelerate its interstellar programme to send a probe there as soon as possible.

The challenges facing the probe are unrivalled. Firstly, it will have to survive for decades in space. Secondly, it will have to be artificially intelligent. It is no use sending a probe which arrives at the star system and sends back a message saying 'What shall I do now?' That message will take years to reach Earth and then years for a reply to be sent back. It is clearly an unworkable situation. Instead, the probe will be virtually a life-form in its own right, with a sophisticated central processing unit that is capable of setting its own mission goals. It will have to be programmed with our knowledge of planetary science and exobiology so that it can prioritise its investigations and carry them out in a logical sequence of events. Mission controllers simply have to sit back and let the probe do all the hard work.

What happens, however, if the probe finds that the star system is inhabited by intelligent beings? Would *we* like it if an interstellar probe suddenly appeared in our Solar System tomorrow and started poking around? One of the greatest science fiction stories every written is Arthur C. Clarke's *Rendezvous with Rama*, in which an alien starship arrives in the Solar System. By the end of the book, when the starship leaves, our heroes and we, the readers, still have no idea what its motives were, if indeed it had any. It was probably just passing through.

Would we programme our probe to hide from intelligent beings, or to say 'hi!' to them? I do not know but, based on the 'listen only' rationale that currently permeates SETI, I suspect we would try to be as undetectable as possible.

Many have suggested that the best way to explore the Galaxy would be to create a spaceprobe capable of building an exact replica of itself. That way, whenever it arrives at a star system, the probe will find the raw materials it needs – probably by mining an asteroid – and build a replica of itself. So where there was one, now there are two. Each goes its separate way, exploring other stars, making a copy of itself on arrival at each new solar system. Each 'generation' doubles the number of such self-replicating probes.

The concept of self-replication was investigated by mathematician John von Neumann in 1953. He discovered that the key to self-replication is a three-part system consisting of a builder, an instruction list and a copier. The builder follows the instruction list and creates the new machine. The final instruction activates the copier. The copier simply copies the instruction list into the new machine, giving it the necessary information to create an exact replica of itself. This is the essence of a von Neumann machine.

Living cells are von Neumann machines. The DNA contains all the information to create new cells and does this by the process of cell division. Effectively a cell grows in size until it splits down the middle, to become two cells. All of the information necessary for the growth is contained in the DNA, which is interpreted by the two-stage transcription and translation process I described in Chapter 4. The final job in cell division, however, is the wholesale copying of the DNA – without any need to know the meaning of genetic code – so that one copy can be retained by the original cell and the second one can be transferred to the new cell. This is exactly von Neumann's approach: build the mechanical part using your instruction list and then program it with a copy of the original instructions.

In terms of a self-replicating spaceprobe, the technology needed to perform such a feat of engineering is almost unbelievable – but 200 years ago the technology to build a television set would have been just as inconceivable. So, I am going to assume that it is possible to construct a von Neumann machine and send it off into the Galaxy at large.

Our self-replicating robot would build a replica of itself and then programme it. This instruction list would not be just the coding necessary to recreate itself but also the basic programming for its artificial intelligence systems and for its mission. If some level of inaccuracy were introduced into the copying, purposely mirroring the occasional genetic mutations that occur in living things, then Darwinian evolution of the robots would become possible. The harsh radiation and the particles that the probe would encounter in space might also inexorably reset or change certain parts of the robot's computerised brain, leading to 'mutations'.

Fatal mutations would cause the probe to malfunction. Advantageous mutations might allow it to finish its survey of a solar system faster than normal and reach the reproduction phase. This would mean that this particular strain of robot would proliferate faster than the others. Perhaps the advantageous trait would be a slight difference in the way it built an instrument scan platform, allowing it to slew between positions faster; or in creating a more efficient way of prioritising its work; or a slightly more efficient propulsion system for travelling between stars. A probe that can potentially improve itself seems great in principle but think about *this* possibility. What if a mutation caused it to 'forget' half its survey mission. This would safely ensure that the robot very quickly began producing more copies of itself. What if the mutation caused it to build *ten* copies of itself, instead of just one?

Suddenly, the release of self-replicating probes capable of setting their own mission parameters does not seem like such a good idea. Machine beings – capable of independent thought, spreading through the Galaxy like a virus, effectively turning into a plague – must be avoided. After just 38 generations – having started with just one probe and assuming that there are no mechanical failures – there would be 275 billion robots – more robots than stars in the Galaxy. Even if a generation were to last 1,000 years and you allow for some of the probes to fail, the conclusion is as chilling as it is inescapable: within a few hundred million years at the very most, the Galaxy would be entirely overrun with machine beings. Biological beings might remain the masters of the planets but the interstellar void would be the home of the machines. Like spawning salmon, they would only return to the confines of a solar system to reproduce and recharge their own energy reserves.

A few hundred million years sounds like a long time but, when we consider that the Galaxy has probably been in existence for 10 billion years, it becomes almost the blink of an eye. At any stage during the preceding 9.9 billion years, an alien civilisation could have released von Neumann machines that should be swarming all over every solar system in the Galaxy, including our own. An obvious question, therefore, is: 'where are they?'

THE FERMI PARADOX

The previous section's analysis, which leads to the question 'where are they?' is known as the Fermi Paradox: if there are aliens, why are they not here? In fact, the argument can worsen if you assume that intelligent species can colonise one planet and then migrate to another one, colonise that and then move on again. Even if it takes 10,000 years to entirely populate a planet which then sends out migration ships, we should be overrun by aliens themselves, not just by self-replicating robots. Despite this reasoning, there is no real, tangible evidence of aliens ever having visited Earth, let alone having participated in its global colonisation.

Sceptics argue that the fact we are not overrun must mean there are no other intelligent beings in the Galaxy. Humankind, they insist, is alone. This forms the central backbone of the space travel argument against extra-terrestrials in John Barrow and Frank Tipler's fascinating tome, *The Anthropic Cosmological Principle*. The fact that the development of intelligence is rare forms the central premise of Brandon Carter's objections in the original Anthropic Principle – out of which grew Barrow and Tipler's work.

The lack of evidence for extraterrestrials is also called the Great Silence by some. Well, for a start, how do we know absolutely for certain that one of the small, so far undetected, asteroids is not an alien probe, quietly watching us? If anyone can prove to me that there is not an extraterrestrial presence in the Solar System at the current time, I will take my inspiration from Paul Horowitz and jump off the radio dish at the University of Hertfordshire's observatory at Bayfordbury. Do not be too impressed, however; Bayfordbury's dish is only six metres in diameter and it is not on a tall mounting. Horowitz put his life on the line but I'm only betting a sprained ankle at worst.

Perhaps they *have* been here. Perhaps they disliked what they saw and left again. Perhaps we live in a zoo or a laboratory, or perhaps there is a *Star Trek* Prime Directive at work, as I mentioned in the previous chapter. If a Galactic civilisation *does* exist, perhaps they have an expansion protocol that forbids the indiscriminate population of other worlds. Perhaps life on other worlds is so different from us that Earth is their idea of a poisonous hell. They might not like space travel because it makes them sick. The list of arguments against the Fermi Paradox – or, if you like, the list of excuses why the aliens are not yet here – is very large and I am sure you can think of some of your own.

Within the last few years, astronomers on Earth have identified the optical explosions associated with gamma-ray bursts in galaxies very far away from ours. Gamma-ray bursts are the most powerful events in the Universe and release so much gamma radiation that they may be able to sterilise the entire galaxy in which they occur. Could a sterilisation event have taken place in the Milky Way from which we are only now just recovering? Looking at it this way, humans could be near to the top of the technological advancement ladder. Maybe there was no sterilisation event and we just happen to be one of the most advanced intelligences in the Galaxy. Who knows what the truth really is? Determining solutions to the Great Silence is immense fun.

Perhaps the most effective argument against the Fermi Paradox has been presented by Jean Heidmann, of the Observatoire de Paris. Heidmann states that any assumption must have a probability of being correct. If a conclusion is based on two assumptions that are each 90% certain, then the conclusion is only 81% certain because $0.9 \times 0.9 = 0.81$. Heidmann has counted 112 assumptions in Barrow and Tipler's space travel argument in *The Anthropic Cosmological Principle* that lead to their conclusion that aliens cannot exist. If all of these 112 assumptions are 90% certain, then the conclusion is 0.9 multiplied by itself 112 times – 0.0007%. Therefore, the conclusion that the Fermi Paradox means that aliens do not exist is only 0.0007% certain. Even if the number of assumptions are reduced, the probability that this argument proves conclusively that aliens do not exist is very small. So, the field is still wide open; the game is still on.

Personally, I think intelligent extraterrestrials are out there and that one day we will find them. What will they be like? What will they breathe? What will they believe in? Will they buy my books? Will they like Rush? We just do not know but every time a biochemist or biologist tries to determine how life forms, every time a SETI astronomer listens to the stars, every time a geologist studies a black smoker, every time an engineer builds a martian rover, we, as a race, are searching for extraterrestrial life and intelligence. At the turn of the third millennium, humankind is like a wolf howling into the lonely night, desperately seeking confirmation that we are not alone. Only by the continued efforts and collaboration of the scientists, who approach the quest from their own unique vantage points, can we hope to find the answer to this, the most perplexing question of our existence, the ultimate test for echo: is anybody out there?

Glossary

α Centauri   The star system nearest to the Sun.

Absorption spectrum   A continuous spectrum on which a number of dark lines are superimposed. It is produced when radiation of all wavelengths passes through a cold gas cloud. The dark lines correspond to absorption at certain wavelengths and the pattern of lines can be used to determine the chemical elements present in the cloud.

ALH84001   A meteorite discovered in the Allen Hills, Antarctica, in 1984. Analysis of trapped gas in the meteorite showed that it originated on Mars. The rock was subsequently discovered to contain evidence that suggested microbial life may once have existed on Mars.

Amino acids   The building blocks of proteins. There are more than 170 known amino acids, only 20 of which are commonly occurring in proteins.

Ammonia (NH$_4$)   One of the astronomical ices. It is a commonly occurring molecule in the Universe.

Anthropic Cosmological Principle   A set of interrelated ideas that arose from the original Anthropic Principle. Readers are directed to the fascinating tome, _The Anthropic Cosmological Principle_, by John D. Barrow and Frank J. Tipler. Some scientists believe that the Principle's ideas are fundamental to nature but others believe that it is simply an exercise in circular argument.

Anthropic Principle   A caveat to the Copernican Principle, proposed by Brandon Carter in 1974. It says that the Earth is privileged, in the sense that it allows the existence of beings capable of observing the Universe. The Principle implicitly assumes that life and intelligence are not common throughout the Universe.

Anticodon   A triplet of three bases found on transfer RNA. These are bound to the triplet of bases, known as a codon, found on messenger RNA. This bonding takes place during the translation stage of protein synthesis and ensures that the correct amino acid is placed in the growing polypeptide chain.

Antimatter   The 'mirror image' of matter. It is opposite in electrical charge but identical in mass, to its matter counterpart. Antimatter is usually produced in energetic reactions that create matter particles. When matter meets its antimatter counterpart, they annihilate each other and release energy.

Artificial intelligence   The abilities displayed by sophisticated computers that are capable of reasoning and judgement.

Asteroid   A relatively small rocky body in the Solar System; a remnant from the formation of the planets. Most asteroids are found in orbits between the orbits of Mars and Jupiter.

Astronomy   The study of the Universe.

Atoms   The smallest piece of matter that can retain a chemical identity.

Attractor   An esoteric concept that pulls a non-linear system into a pattern of repeated behaviour.

Barycentre   The point between a system of two gravitating bodies where their gravitational attraction cancels out. Also called the 'centre of mass'.

Base   In the context of DNA, a base is a nitrogen-rich organic molecule that can bond to another, specific base. Adenine bonds to thymine, and cytosine bonds to guanine. The bonding between bases on two DNA strands pulls it into the famous double-helix structure. In RNA, thymine is replaced by the base uracil.

Big Bang   A theory that explains the origin of the Universe.

Black smoker   A hydrothermal vent on the ocean floor. The name derives from the precipitation of chemicals that produces clouds of black 'smoke' around the vents.

Book of Life project   An endeavour by scientists at NASA's Marshall Space Flight Center to train a neural network computer to recognise microbial life forms.

Bugs from space   An alternative name for the 'panspermia' hypothesis.

Bussard ramjet   A hypothetical device for collecting hydrogen gas from interstellar space. It would be used by a starship to gather fuel for its engine – a nuclear fusion reactor.

Cambrian explosion   The sudden diversification of species 535 million years ago.

Carbon dioxide (CO$_2$)   One of the astronomical ices; an abundant molecule in the Universe.

Cell   The smallest living structure known to humankind. A cell contains the genetic information necessary for self-replication and all the machinery necessary for protein synthesis.

Cerebellum   The part of the brain that co-ordinates muscular movements, balance and co-ordination.

Cerebral cortex   The wrinkled surface of the cerebrum. This is the seat of intelligence, defined as the ability to solve problems.

Cerebrum   The part of the brain that processes sensory input. It is divided into two hemispheres.

Chaos theory   The collection of ideas that allows scientists to understand systems that rapidly diverge from predictions because they are extremely sensitive to initial conditions and even tiny perturbations.

Charge-coupled device (CCD)   A type of digital camera that, in astronomy, has virtually replaced photographic film for recording images of the night sky. The image information can be directly downloaded to computers.

Chert   A type of sedimentary rock.

CHNOPS elements   The chemical elements that are essential for Earth life: carbon, hydrogen, nitrogen, oxygen, phosphorus and sulphur.

Clockwork Universe   A belief that the Universe was totally predictable. This idea sprang out of Newton's laws of motion that could predict movement both on Earth and in the heavens. Chaos theory is the ultimate disproof of the clockwork Universe.

Closed system   Any system in which energy cannot enter or leave.

Codons   Sequences of three bases on either DNA or RNA strands that code for specific amino acids.

Comet   A relatively small icy body, left over from the formation of the Solar System. Most orbit the Sun way out beyond Pluto. Periodically, some pass through the inner Solar System.

Contingency   The concept that each evolutionary step is contingent upon the previous step. Hence, a tiny change in one step should cause the organism to evolve in a completely different direction. This concept is partially refuted by that of convergent evolution.

Continuous spectrum   A rainbow spectrum of colours produced by a hot object. The wavelength at which most energy is being released can be used to deduce the temperature of the emitting object.

Convergent evolution   The observational fact that sometimes organisms grow similar structures from completely different evolutionary pathways in order to combat common environmental problems. An example is the development of the eye in 40 distinct species.

Copernican Principle   This principle states that there is nothing special about the Earth or its location in space.

Corpus callosum   A band of fibres that connect the hemispheres of the cerebrum. The corpus callosum is the part of the brain responsible for animalistic sounds and utterances.

Cosmic Background Explorer (COBE)   A highly successful NASA satellite, launched in 1989. It discovered the seeds of cosmic structure, imprinted on the cosmic microwave background radiation.

Cosmic microwave background radiation   Relic radiation from the Big Bang that floods the Universe. There are approximately one billion photons of this radiation for every particle of matter.

Cosmicrobia   An alternative term for 'panspermia'.

Cosmology   The study of our Universe's origin.

Cynobacteria   Single-celled organisms that metabolise carbon dioxide rather than oxygen.

Cytoplasm   The contents of a eukaryote cell contained between the nucleus and the cell membrane.

Darwin project   A European Space Agency project to place in space a number of telescopes that will be capable of searching for Earth-like worlds around other stars.

Decoupling of matter and energy   The point in time, 300,000 years after the Big Bang, when neutral hydrogen atoms formed for the first time and the Universe became transparent to radiation.

Deep hot biosphere   A theory, proposed by Thomas Gold, that deep inside the rocks are vast numbers of single-celled creatures.

Deuterium   A form of hydrogen that contains one neutron in its nucleus, as well as a proton.

Dissipative structure   An open thermodynamic system that is capable of organising itself because of the flux of matter or energy running through it.

DNA (deoxyribonucleic acid)   The information-carrying molecule that makes life on Earth possible.

Doppler effect   The stretching or squashing of waves because of movement between the source and the observer.

Dyson sphere   A hypothetical construction that surrounds a star with space-habitats so that the entire radiation output of a star can be used by a civilisation.

Ediacara   A group of creatures that lived just before the Cambrian explosion. All species of Ediacara are now extinct.

Electron   The naturally occurring negatively charged particle of nature.

Emergence   The process whereby traits, manifested by a system, could not be predicted from a knowledge of the system's constituents.

Emission spectrum A characteristic pattern of coloured lines that are emitted by a hot gas. An analysis of the pattern of emission lines will reveal the chemical identity of the emitting gas.

Energy A property required in order to perform work.

Entropy A measure of the disorganisation of a thermodynamic system.

Enzyme An organic molecule that is capable of making a reaction proceed faster than it would without the presence of the enzyme.

Eukaryotes A cell with a nucleus in which DNA is found.

Europa Orbiter A NASA mission that is planned to map Europa towards the end of the first decade of the third millennium

Exobiology The study of life in locations other than Earth.

Exploration by data exchange An assumption that alien civilisations will exchange knowledge of the Universe via radio waves.

Extrasolar planets Planets orbiting other stars.

Extraterrestrials Alien beings.

Extremophiles Organisms that are capable of surviving extreme conditions that would kill other Earth life.

Fermi Paradox The statement that if extraterrestrials exists, they should have visited Earth. Since we see no evidence of their visitation, they cannot exist.

Galactic Club The assumption that alien civilisations form a 'club' that communicates via radio waves.

Galaxy A collection of, typically, a few hundred billion stars.

Galilean satellites The four moons of Jupiter (Io, Europa, Ganymede and Callisto) that were discovered by Galileo.

Galileo spaceprobe A NASA spaceprobe that successfully explored Jupiter and its moons in the late 1990s.

Gamma-ray bursts Incredibly powerful events that take place at random throughout the Universe. Most are at extreme distance and are thought to be linked to exploding stars.

General intelligence factor A number that supposedly quantifies the intelligence of an individual. Many psychologists are sceptical of this simplistic approach to intelligence.

Genes Sections of DNA that contain the building instructions for individual proteins.

Gravitons Hypothetical particles that carry the force of gravity. As yet, no convincing theory exists to explain them.

Gravity The attractive force that pulls together all objects with mass. It is one of the four fundamental forces of the Universe.

Great Red Spot   An enormous weather system in the atmosphere of Jupiter. It has persisted for centuries.

Great Silence   The term used to describe the fact that SETI searches have so far detected nothing.

Greenhouse effect   The effect whereby greenhouse gases, such as water and carbon dioxide, trap heat in the atmosphere of a planet, raising the surface temperature.

Grey matter   The 2-mm thick layer of cells in the cerebral cortex.

Gusev crater   A feature on Mars; possibly a dried-up water lake.

Habitable zone   The region around a star within which a planet can possess liquid water due to stellar heating. The exact boundaries of the zone depend on the nature of the star and the nature of the planet's atmosphere.

Heavy bombardment phase   The final stage in the formation of our Solar System in which left-over comets and asteroids pelted the planets. This was the stage when most of the volatile substances, such as water, were brought to the inner Solar System worlds.

Helium 3   A light form of helium that possesses only one neutron and two protons in its nucleus.

High Resolution Microwave Survey (HRMS)   The SETI search, developed by NASA's Ames Research Center, that was cancelled by the US Congress. This search went on, to be resurrected by the SETI Institute as Project Phoenix.

Hominids   The collective term for humans and their cavemen ancestors.

Hox cluster   A collection of genes that allow a developing creature to grow body parts in the right place – a head at the front, legs underneath, wings on top and so on.

Hubble Deep Field   An unprecedented 10-day exposure of deep space that resulted in the discovery of 3,000 new galaxies.

Huygens probe   Part of the Cassini–Huygens mission to Saturn. In 2004, Huygens will land on Titan, Saturn's largest moon.

Hydrobots   Conceptual spaceprobes, designed to explore the subsurface ocean of Europa.

Hydrothermal vents   Surface features that release volcanically heated water. On the ocean floor they can lead to the production of black smokers.

Hyperthermophile   A micro-organism capable of living in extremely hot water.

Inevitability of life and technology   An assumption made by some SETI scientists that everywhere life develops, it will evolve to be intelligent and use technology such as radio telescopes.

Inflation   A theory that the Universe underwent a period of accelerated expansion shortly after its birth.

Intelligence Quotient (IQ)   A hotly-debated method for quantifying the intelligence of an individual.

International Astronomical Union   A governing body that sets guidelines and standards for the astronomical community. They also officially name celestial objects.

Isotopes   Chemical elements that have different numbers of neutrons from normal in their atomic nuclei.

Junk DNA   Apparently meaningless DNA. It constitutes 85–90% of all DNA in a cell.

Kinetic energy   Energy expressed as movement.

Lake Vostok   A subsurface lake, the size of Lake Ontario, that has been discovered in Antarctica.

Left lateral language area   The part of the brain in humans that governs speech.

Linear system   A system in which the prediction of future behaviour is relatively easy. Small variations in input result in only small variations in output.

Mass   A property that quantifies how much matter an object contains.

McMurdo Dry River Valleys   An area of Antarctica where ice-covered lakes are found. The average temperature is −68° C and fewer than four inches of rain fall every year.

Medulla oblongata   The area at the top of the spinal column that joins the brain. The *medulla oblongata* controls respiration, heart-beat and blood pressure.

Messenger RNA (mRNA)   Strands of RNA that carry genetic information from the DNA strand to the ribosome so that proteins can be built.

Meteorites   Small pieces of rock that fall to Earth from space.

Methane (CH_3)   One of the astronomical ices, abundant in the Universe.

Microbe   A single-celled organism that is typically a few hundred microns in length.

Milky Way   Our Galaxy. From clear, dark sites, the Milky Way can be seen stretching across the sky.

Miller–Urey experiments   Experiments designed to simulate the conditions of the primordial Earth's atmosphere to determine whether organic molecules can be easily synthesized.

Moon   A small celestial object in orbit around a planet.

Multiple intelligences   A theory that intelligence is multifaceted, rather than a single trait. Each of us is gifted to lesser or greater degrees in each of the intelligences. These include musical intelligence, linguistic intelligence, kinaesthetic intelligence and logical–mathematical intelligence.

Murchison meteorite   A meteorite that was found to contain a variety of amino acids, indicating that these building blocks of protein are synthesized in space before the formation of planets.

Mutations   The process by which genes do not copy exactly. Mutations lead to differences in the construction of proteins. Some mutations provide life-forms with advantages but the vast majority do not.

Nakhla meteorite   A meteorite that has been shown to come from Mars and which may contain evidence of past martian life.

Nanobes   The latest name given to a group of living organisms that are so small that they are measured in hundreds of nanometres.

Neural network   A computer with an operating technique which mimics some aspects of the working of the human brain.

Neutrons   Neutral particles of matter. Neutrons dictate the isotope of an element.

Newton's three laws of motion   1: Bodies remain at rest or continue in a state of uniform motion unless acted upon by an external force. 2: The force exerted on a body is equal to the body's mass multiplied by the acceleration that the force causes. 3: For every action there is an equal and opposite reaction.

Non-linear system   A system that rapidly diverges from predictions. Small variations in input can result in large variations in output.

No superscience or hyperdrives   An assumption of some SETI scientists that restrict the technological development of extraterrestrials so that they are forced to rely on radio waves to communicate across interstellar distances.

Nucleotide   The repeated component of a nucleic acid. In the case of DNA, a nucleotide is made up of a sugar consisting of five carbon atoms, a phosphate acid and a nitrogen-rich base.

Nulling interferometer   A system of two of more telescopes that can be operated together to reduce the glare of a central bright object, such as a star.

Olympus Mons   A volcano on Mars; the largest volcano in the Solar System.

One-hectare Telescope, 1hT   The name of a newly-proposed project, by the University of California at Berkeley, to build a new radio telescope, substantially dedicated to SETI.

Open thermodynamic system   Any system capable of taking in matter or energy from its surroundings and expelling a waste product.

Organelle   A specialised piece of organic machinery that exists within an eukaryote cell and performs a specific biological function.

Organic chemistry   The branch of chemistry that concerns itself with carbon compounds.

Organism   A living thing.

Panspermia   The theory that life did not originate on Earth but was brought here as microbes protected from the harsh radiation environment of space.

Perfect gas   A gas in which every particle behaves like a billiard ball.

Photon   A particle of electromagnetic energy.

Planet   A relatively large celestial body in orbit around a star.

Planetary nebula   A nebula created when a low-mass star like the Sun reaches the end of its energy-generating life.

Planetesimal   An asteroid-like body created in the first stage of planetary formation.

Plate tectonics   The theory that explains the way in which the continents on Earth move relative to each other. It relies on the existence of a molten layer of rock beneath the Earth's surface.

Polypeptide chain   A molecule made of three or more amino acids. A protein may consist of one or more polypeptide chains.

Potential energy   The stored energy an object contains because it is suspended in the gravitational field of a large object.

Primordial soup   The primordial constituency of Earth's oceans.

Principle of Plenitude   Anything that *can* happen, *will* happen.

Principle of Uniformity of Nature   The laws of nature are the same everywhere in the Universe.

Project Cyclops   An ambitious NASA proposal for a SETI programme that would have been sensitive enough to eavesdrop on the stray broadcasts from alien civilisations. It was deemed too expensive to be funded.

Project Phoenix   The resurrected High Resolution Microwave Survey, HRMS, cancelled by the US Congress but revived by the SETI Institute.

Project Ozma   The first, structured SETI ever undertaken on Earth. It was conducted by Frank Drake in 1959.

Prokaryotes   Single-celled creatures that lack a nucleus.

Protein   A general term for essential organic molecules made of amino acids. There are moer than 10,000 different proteins found in living things.

Protons   Positively-charged particles of matter that determine the chemical identity of atoms.

Punctuated equilibrium   An evolutionary theory in which the creation of new species takes places in 'fits and starts' rather than in a constant, gradual process.

Pythagoras' theorem   In a right-angled triangle, the square of the hypotenuse equals the sum of the squares of the other two sides.

Quantum entanglement   Phenomenon whereby collections of particles are linked and can communicate information instantaneously.

Quantum theory   A theory that relates physical effects to the action of subatomic particles.

Redshift   The stretching of light rays to longer wavelengths by a variety of physical mechanisms, including the Doppler effect.

Reductionism   The scientific principle of reducing a problem to its simplest, most understandable parts.

RNA (ribonueleic acid)   A nucleic acid essential for life. It takes part in protein synthesis.

Ribosome   The organelle in a eukaryote cell, responsible for translating the mRNA into proteins.

Ribosomal RNA   The RNA that makes up ribosomes.

RNA polymerase   An enzyme that assists in the transcription of DNA onto mRNA by unwrapping the DNA strand.

Self-organisation   The ability of a thermodynamic system to decrease its entropy and become more organised.

Self-preservation   The ability of a life-form to keep itself alive.

Self-regulation   The ability of a life-form to keeps itself working at optimum efficiency.

Self-reproduction   The ability of a life-form to reproduce without the assistance of an external agent.

Self-replicating robot   A conceptual robot that is capable of building an exact replica of itself.

SETI (Search for Extraterrestrial Intelligence)   The search for extraterrestrial intelligence by the strategy of trying to detect extraterrestrial radio broadcasts.

SETI@home   A screen-saver program that allows home computers to analyse data from project SEREDIP, in an attempt to detect extraterrestrial messages.

Solar system   A collection of planets orbiting a star.

Space-travel argument   Used in the Anthropic Cosmological Principle to argue against the existence of intelligent extraterrestrials. It is basically a rewording of the Fermi Paradox.

Spectroscope   A device capable of splitting light into a spectrum.

Sputtering   The gradual erosion of a planetary atmosphere by the influx of high-energy particles.

Star A ball of gas capable of nuclear fusion; or, the remains of a body that was once capable of nuclear fusion.

Strange attractor An esoteric concept that dictates the sudden emergence and destruction of order in a chaotic system.

Stromatolites Large-scale structures that are built up by the action and growth of colonies of bacteria.

Sublimation A phase change in which a solid turns immediately into a gas.

Temporal mediocrity States that the current time is no more important than any other time in the Universe for the existence of life-forms.

Terrestrial Planet Finder A NASA design for a flotilla of space telescopes that would be capable of imaging Earth-like worlds orbiting other stars.

Theory of general relativity Albert Einstein's famous theory in which he provides science with a mechanism for the understanding of gravity.

Thermal equilibrium A situation in which two or more bodies are at the same temperature.

Thermalisation The process by which objects are brought into thermal equilibrium.

Thermodynamics The study of heat and energy flow.

Thermodynamics, first law Energy can neither be created nor destroyed.

Thermodynamics, second law Heat flows from hot bodies to cold bodies.

Transcription The first stage in the synthesis of proteins. The genetic code for the protein is transcribed from the DNA onto a strand of mRNA.

Translation The second stage in the synthesis of proteins. The genetic code on the mRNA is used to assemble a polypeptide chain consisting of amino acids.

Tritium A form of heavy hydrogen in which the nucleus contains a proton and two neutrons.

tRNA Transfer RNA exists in the cytoplasm and is used to transport amino acids into their correct position on a growing polypeptide chain during the process of protein synthesis.

Tufas Sedimentary mineral structures found in lakes with a high mineral content.

Universe Everything that there is, was, or ever will be.

Viking spaceprobes A pair of successful NASA spaceprobes that landed on Mars in the mid-1970s.

Vitalism The outdated belief that life can be created only if inorganic matter is suffused by some mysterious vital force.

Water (H_2O) One of the astronomical ices. Water is essential for life on Earth and may be universally essential for life. It is a molecule abundant in the Universe.

Waterhole A 300 MHz band of radio frequencies bounded by emission lines of hydrogen (H) and hydroxyl (OH). This waveband exists in a part of the electromagnetic spectrum where background interference drops to a minimum.

Wave–particle duality A cornerstone of the quantum theory. It states that, under certain conditions, waves can behave like particles and particles can behave like waves.

White matter The fibrous connections between the grey matter of the cerebral cortex and the rest of the brain.

Bibliography

Books

Ashpole, E., *The Search for Extraterrestrial Intelligence*, Blandford, 1989

Barrow, J., *The Artful Universe*, Penguin, 1995

Barrow, J. and Tipler, F.J., *The Anthropic Cosmological Principle*, Oxford, 1986

Cairns-Smith, A.G., *Seven Clues to the Origin of Life*, Cambridge, 1995

Calvin, W.H., *How Brains Think*, Phoenix, 1996

Chown, M., *Afterglow of Creation*, Arrow, 1993

Chown, M., *The Magic Furnace*, Jonathan Cape, 1999

Clark, S., *Extrasolar Planets*, Wiley–Praxis, 1998

Clark, S., *Towards the Edge of the Universe*, Springer–Praxis, 1999

Crowe, M.J., *The Extraterrestrial Debate 1750–1900*, Cambridge, 1986

Davies, P., *Are we Alone?*, Penguin, 1995

Davies, P., *The Last Three Minutes,* Basic Books, 1994

Finny, B.R. and Jones, E.M., (*eds.*), *Interstellar Migration and the Human Experience*, University of California Press, 1986

Gold, T., *The Deep Hot Biosphere*, Copernicus, 1999

Hagene, B. and Lenay, C., *The Origin of Life*, Barron's, 1987

Heidmann, J., *Extraterrestrial Intelligence*, Cambridge, 1997

Holland, J.H., *Emergence: From Chaos to Order,* Oxford, 1998

Jakosky, B., *The Search for Life on Other Planets*, Cambridge, 1998

Kane, G., *The Particle Garden*, Addison–Wesley, 1995

Kauffman, S., *At Home in the Universe*, Penguin, 1995

Mallove, E. and Matloff, G., *The Starflight Handbook*, Wiley, 1989

Miller, S.L. and Orgel, L.E., *The Origins of Life on Earth*, Prentice-Hall, 1974

Ponnamperuma, C. and Cameron, A.G.W., *Interstellar Communication: Scientific Perspectives*, Houghton Mifflin Company, 1974
Press, F. and Raymond, S., *Earth*, Freeman, 1986
Purves, B., Heller, C., Orians, G. and Sadava, D., *Life*, Freeman, 1997
Reeves, H., *The Hour of Our Delight*, Freeman, 1991
Regis Jr, E. (*ed.*), *Extraterrestrials: Science and Alien Intelligence*, Cambridge, 1987
Schrödinger, E., *What is Life?*, Cambridge, 1998
Stewart, I., *Life's Other Secret*, Penguin, 1998
Tovmasyan, G.M., *Extraterrestrial Civilisations*, Ann Arbor-Humphrey, 1967
Zuckerman, B. and Hart, M.H. (*eds.*), *Extraterrestrials: Where Are They?*, Cambridge, 1995

Magazines

Clark, S., Polarized Starlight and the Handedness of Life, *American Scientist*, **87**, no. 4, July–August 1999
Exploring Intelligence, *Scientific American Presents*, **9**, no. 4, Winter 1998
Millennium Special: Other Worlds, *New Scientist*, **163**, no. 2204
The Future of Space Exploration, *Scientific American Presents*, **10**, no. 1, Spring 1999

Index